The Vascular Plants of British Columbia

Part 4 – Monocotyledons

edited by

George W. Douglas[1], Gerald B. Straley[2] and Del Meidinger[3]

[1] Douglas Ecological Consultants Ltd.
6200 North Road
R.R. #2
Duncan, B.C. V9L 1N9

[2] Botanical Garden
University of British Columbia
6501 N.W. Marine Drive
Vancouver, B.C. V6T 1W5

[3] Research Branch
B.C. Ministry of Forests
31 Bastion Square
Victoria, B.C. V8W 3E7

Ministry of Forests
Research Program

1994

Canadian Cataloguing in Publication Data

Main entry under title:
The Vascular plants of British Columbia

(Special report series, ISSN 0843-6452 ; no. 4)

"With the cooperation of the Royal British Columbia Museum and the Botanical Garden, University of British Columbia."
Includes bibliographies and indexes.
Contents: Pt. 4. Monocotyledons.
ISBN 0-7718-8757-4 (set). -- ISBN 0-7718-8761-2 (pt. 4)

1. Gymnosperms. 2. Dicotyledons - British Columbia. 3. Pteridophyta - British Columbia. 4. Monocotyledons. 5. Botany - British Columbia. I. Douglas, George W., 1938- . II. Straley, Gerald Bane, 1945- . III. Meidinger, Dellis Vern, 1953- . IV. British Columbia. Ministry of Forests. V. Royal British Columbia Museum. VI. University of British Columbia. Botanical Garden. VII. Title: Monocotyledons. VIII. Series: Special report series (British Columbia. Ministry of Forests) ; no. 4.

QK203.B7V37 1989 581.9711 C89-092095-8

Published by the
Research Branch
Ministry of Forests
31 Bastion Square
Victoria, B.C. V8W 3E7

Copies of this and other Ministry of Forests
titles are available from
Crown Publications Inc.
546 Yates Street
Victoria, B.C. V8W 1K8

THE VASCULAR PLANTS OF BRITISH COLUMBIA

edited by

George W. Douglas, Gerald B. Straley and Del Meidinger

Part 4 – Monocotyledons

Contributors:

Dr. S.G. Aiken, Research Division, Canadian Museum of Nature, Ottawa, ON K1A 0M8 – *Festuca*

Dr. M.E. Barkworth, Intermountain Herbarium, UMC 45, Utah State University, Logan, UT 84322-5305 – *Agropyron, x Elymordeum, Elymus, Hordeum, Glyceria, Leymus, Melica, Nassella, Oryzopsis, Pleuropogon, Secale, Stipa, Tricetum*

Ms. E.M. Campbell, Research Branch, Ministry of Forests, 31 Bastion Square, Victoria, BC V8W 3E7 – Glossary

Dr. A. Ceska, Botany Division, Royal British Columbia Museum, Victoria, BC V8V 1X4 – *Carex*, Juncaceae, Potamogetonaceae

Dr. W.J. Crins, Biological Survey, New York State Museum, University of the State of New York, Albany, NY 12230 – *Alopecurus, Ventenata.* Present address: Ontario Ministry of the Natural Resources, Box 9000, Huntsville, ON POA 1KO

Dr. S.J. Darbyshire, Centre for Land and Biological Research, Agriculture Canada, Central Experimental Farm, Canadian Dept. of Agriculture, Ottawa, ON K1A 0C6 – *Danthonia*

Dr. J.I. Davis, L.H. Bailey Hortorium, Cornell University, Ithaca, NY 14853-4301 – *Puccinellia, Torreyochloa*

Dr. G.W. Douglas, Douglas Ecological Consultants Ltd., 6200 North Road, R.R. 2, Duncan, BC V9L 1N9 – Alismataceae through Hydrocharitaceae, Juncaginaceae through Lemnaceae, Najadaceae, Poaceae, Pontederiaceae and Ruppiaceae through Zosteraceae

Dr. C.W. Greene, Faculty in Biology, College of the Atlantic, Bar Harbour, MA 04609 – *Calamagrostis*

Dr. M.J. Harvey, Dept. of Biology, Dalhousie University, Halifax, NS B3H 4J1 – *Agrostis.* Present address: 5061 Sooke Road, R.R. 2, Victoria, BC VOS 1NO

Dr. R. Lonard, Dept. of Biology, Pan American University, Edinburg, TX 78539 – *Vulpia*

Dr. J.G. Packer, Department of Botany, University of Alberta, Edmonton, AB T6G 2E9 – *Panicum*

Mr. L.E. Pavlick, Botany Division, Royal British Columbia Museum, Victoria, BC V8V 1X4 – *Festuca*

Dr. P.M. Peterson, Dept. of Botany, National Museum of Natural History, Smithsonian Institution, Washington, DC 20560 – *Eragrostis, Muhlenbergia*

Dr. R.J. Soreng, Biology Dept., Virginia Polytechnic & State University, Blacksburg, VA 24061 – *Poa.* Present address: L.H. Bailey Hortorium, Cornell University, Ithaca, NY 14853

Dr. G.B. Straley, Botanical Garden, University of British Columbia, Vancouver, BC V6T 1Z4 – Iridaceae, Liliaceae, Orchidaceae

With the cooperation of the Royal British Columbia Museum and the Botanical Garden, University of British Columbia.

ACKNOWLEDGEMENTS

We are grateful to Dr. G.A. Allen for providing valuable suggestions during the initial stages of the project.

Thanks are also due to Ms. A. Nicholson, Mr. L. Pavlick and Drs. D.F. Murray, S.G. Aiken, S.J. Darbyshire, M. Barkworth, and R.R. Haynes for reviewing taxonomically difficult groups. Dr. A. Ceska and Mr. L. Pavlick also provided valuable comments on several taxa.

Ms. A. Roberts, and Drs. T.C. Brayshaw, W.J. Crins, H. Roemer, J. Pojar and P. Warrington kindly reviewed the entire manuscript.

Louise Gronmyr, Christina Stewart, and Trish Simoes typed many of the contributions and helped in many ways in the production of the final manuscript. We also appreciate the work of T.D. Mock & Associates, especially Christine Smith, for dealing with the final formatting of the document and preparation of the index.

We are thankful to the curators and staff of the following institutions for allowing us to study their collections and use their facilities: National Herbarium, National Museums, Ottawa (CAN); Herbarium, Royal British Columbia Museum, Victoria (V); Vascular Plant Herbarium, Department of Agriculture, Ottawa (DAO); Herbarium, Department of Botany, University of British Columbia, Vancouver (UBC); Herbarium, Department of Biology, University of Victoria (UVIC); Herbarium, Department of Botany, University of Toronto (TRT).

We would also like to thank and congratulate those botanists who collected new BC records during the past few years. These botanists and their discoveries include: P. Achuff (*Agoseris lackschewitzii*), B. Bennett (*Petrohagia prolifera)*, A. Ceska (*Coleanthus subtilus, Floerkea proserpinoides* and *Scirpus supinus* var. *saximontanus*), J. Cumming (*Hesperochiron pumilus)*, T. Goward (*Astragalus spaldingii* var. *spaldingii*), F. Lomer (*Cnidium cnidiifolium, Deschampsia flexuosa, Eriophorum virginicum, Leptochloa fascicularis, Rumex patienta* and *Sporobolus vaginiflorus*), M. Martin (*Chondrilla juncea, Polycnemum arvense* and *Rumex patienta*), H. Moore (*Phacelia ramosissima*), J. Pojar and D. Meidinger (*Chrysosplenium wrightii* and *Saxifraga davurica* ssp. *grandipetala*) and W.F. Savale (*Wolfia columbiana*).

TABLE OF CONTENTS

INTRODUCTION

This manual describes all of the flowering plants and vascular cryptogams, both native and naturalized, occurring within the borders of British Columbia. It includes keys to genera, species, subspecies and varieties and a glossary. The study is based on examination of specimens in all major Canadian herbaria and some U.S. herbaria (e.g., UC, CAS, WTU), as well as a complete review of the pertinent literature.

Species not represented by a herbarium collection have been placed in an excluded list (see Appendix). Many of these were incorrectly assumed, by American authors, to range into Canada. These species were often taken up by Taylor and MacBryde (1977). We have also excluded many introduced species that are known from only a few collections taken more than 40 years ago.

FORMAT

The manual has been published in four parts – Part 1 contains Gymnosperms and Dicotyledons [Aceraceae through Cucurbitaceae]; Part 2 contains Dicotyledons [Diapensiaceae through Portulacaceae]); Part 3 contains Dicotyledons [Primulaceae through Zygophyllaceae] and Pteridophytes, and Part 4 contains the Monocotyledons. We plan to publish the entire manual, with illustrations and technical descriptions in two volumes.

The floristic treatment is in alphabetical order by families, genera and species since most of our users may not be familiar with the more traditional "Englerian system". Expected users of this treatment (e.g., foresters, wildlife specialists, park managers, amateur naturalists, students, etc.) often require specific information about a species. Thus we have attempted to supply the most frequently required information including:

Synonyms:

Synonyms are given when there are references in major floristic treatments to a given species by another name or names.

Common Names:

A common name for each species is included. These are based mainly on names appearing most commonly in the Western North American literature.

Habitat and Moisture Regime:

A brief description of a species habitat preference, including moisture regime, is given. The latter uses a simple (wet – moist – mesic – dry) scale.

Vegetation Zones:

The elevational limits of a species are presented using vegetation zones. The zonal system used here, unlike some of the more localized zonation systems (Krajina 1965; Rowe 1959, 1972; Franklin and Dyrness 1973), has wide application in the northern hemisphere and differs only slightly from that proposed by Danserau (1975) and Meusel *et al.* (1965) and recommended by Löve (1970). The montane zone includes all continuous forests in British Columbia except for the coastal lowlands and some islands which are included in the lowland zone. The upper forests in southern British Columbia, included by some local ecologists (e.g., Krajina 1965) in the subalpine zone, is considered the upper montane zone in this work. The subalpine zone is defined as that area above the montane zone and below the upper limit of conifers as an upright tree form (Douglas 1972). In the southern part of the province, subalpine vegetation consists of a meadow/tree-clump complex (Douglas 1971, 1972), while in the northern part of the province the subalpine zone is dominated by tall shrubs (mainly three- to four-metre *Salix*) and scattered trees (Douglas 1974, Krajina 1975).

Above the subalpine is the alpine zone where trees occur only in krummholz (or dwarfed) form and the vegetation is extremely short (less than one metre) and commonly referred to as alpine tundra (Krajina 1969, Douglas 1972, Douglas and Bliss 1977). The steppe vegetation zone (Daubenmire 1970) occurs in the interior of the province and includes what is often referred to as sagebrush or grassland vegetation. The former occurs at lower elevations in the Chilcotin, Thompson, Okanagan, Kootenay, Columbia, and Flathead River valleys in southern British Columbia while the latter includes parts of the Fraser and Thompson plateaus and upper Peace River drainage.

Abundance:

The abundance of a species is given by a general scale: rare – infrequent – frequent – common. The rare, and some of the infrequent plants, are well documented (at least to the best of our knowledge) in Straley *et al.* (1985). Most of the common plants are also well known from numerous other sources. In several cases, however, limited collecting or research enabled us to make only general assumptions about the abundance of some of the infrequent or frequent species.

Range:

The general distribution of a species, both within and outside the province, is presented. Maps compiled at the Royal British Columbia Museum were extremely helpful. Abbreviations are as follows:

Alberta	AB	Maine	ME
British Columbia	BC	Maryland	MD
Manitoba	MB	Massachusetts	MA
New Brunswick	NB	Michigan	MI
Newfoundland	NF	Minnesota	MN
Northwest Territories	NT	Mississippi	MS
Nova Scotia	NS	Missouri	MO
Ontario	ON	Montana	MT
Prince Edward Island	PE	Nebraska	NE
Quebec	PQ	Nevada	NV
Saskatchewan	SK	New Hampshire	NH
Yukon	YT	New Jersey	NJ
		New Mexico	NM
Alabama	AL	New York	NY
Alaska	AK	North Carolina	NC
Arizona	AZ	North Dakota	ND
Arkansas	AR	Ohio	OH
California	CA	Oklahoma	OK
Colorado	CO	Oregon	OR
Connecticut	CT	Pennsylvania	PA
Delaware	DE	Rhode Island	RI
District of Columbia	DC	South Carolina	SC
Florida	FL	South Dakota	SD
Georgia	GA	Tennessee	TN
Hawaii	HI	Texas	TX
Idaho	ID	Utah	UT
Illinois	IL	Vermont	VT
Indiana	IN	Virginia	VA
Iowa	IA	Washington	WA
Kansas	KS	West Virginia	WV
Kentucky	KY	Wisconsin	WI
Louisiana	LA	Wyoming	WY
		Mexico	MX

Notes:

Specific items of importance or problems, especially those related to taxonomy and nomenclature, are noted. If infraspecific taxa require recognition these are keyed here.

A complete floristic bibliography, arranged by families, is also appended for users wishing to study a species further. A glossary also appears in Part 4.

The absence of descriptions and illustrations, due to severe budget limitations, should not deter from the overall treatment. Technical descriptions and illustrations of most of British Columbia's plant species are available in a number of regional floras (Abrams 1923–1951; Ferris 1960; Hitchcock *et al.* 1955–1969) or in treatments of some British Columbian plant families (Brayshaw 1976, 1985, 1990; Douglas 1982, 1993, 1994; Szczawinski 1959, 1962; Taylor 1963, 1966, 1970, 1973a, 1974a, 1974b, 1983). Additional descriptions for the remaining plants may be found in Argus (1973), Fernald (1950), Packer (1983), Porsild and Cody (1980), Tutin *et al.* (1964–1980) or Welsh (1974).

DATA BASE

Historically, the floristic documentation of British Columbia's flowering plants and vascular cryptogams began with the enumeration of the extensive collections of J. Macoun (Macoun 1883–1890). This was soon followed by the botanical contributions of his son, J.M. Macoun (1889, 1894–1906). In 1915, the first provincial flora appeared (Henry 1915), treating the southern part of the province. A supplement to this work was later provided by Eastham (1947). The southern part of the province was also included in a treatment of Pacific Northwest vascular plants by Hitchcock *et al.* (1955–1969) – later edited and revised by Hitchcock and Cronquist (1973). During this same period several other major works also appeared. Boivin (1966–1967), after numerous visits to Canadian and American herbaria, produced a checklist of Canadian vascular plants. In his Flora of Alaska and Neighboring Territories, Hultén (1968) provided brief descriptions with range maps that, in many cases, included British Columbian species. Meanwhile, in a treatment of Queen Charlotte Island vascular plants, Calder and Taylor (1968) contributed one of the few floras dealing solely with British Columbia. British Columbia was again included in a regional flora of northwestern North America when Welsh (1974) treated the plants of northern British Columbia. A literature review by Taylor and MacBryde (1977), mainly of North American studies, resulted in a resource inventory of vascular plants. Unfortunately, this extensive review treated several hundred erroneous citations as valid records for the province. Recently, a modern Canadian flora was produced by Scoggan (1978–1979). Finally, all of the province's rare or infrequent plants were examined and verified in various Canadian herbaria by Straley *et al.* (1985).

In addition to the major works mentioned above, there have been numerous taxonomic, floristic and related research papers dealing with British Columbia's vascular plants. Many of these are cited in the floristic bibliography by Douglas *et al.* (1983) and many are specifically referenced in this manual.

TAXONOMIC CONCEPTS

Our view of species limits is a practical one, with species characterized morphologically in order to produce keys usable by a variety of users. Many of the keys were newly constructed while others were adapted for our British Columbia plants from well-tested manuals.

Wherever possible, geographically separate, or largely separate, infraspecific taxa are treated at the subspecies level, while those with geographically sympatric ranges are treated at the varietal level. Since we have not made any nomenclatural changes associated with the production of the manual, this choice between subspecies and variety was not always possible. Only one infraspecific rank is used for a species, and formae are not recognized.

PHYTOGEOGRAPHY

The native vascular plant flora of British Columbia has a complex and varied history, especially with respect to glaciation. It consists of a total of 2301 taxa within a number of floristic elements and subelements. These elements range from widespread cosmopolitan species to restricted local endemics. The synopsis presented here includes the major elements in the flora, some of their subelements, the number of native BC taxa within these and examples of representative taxa.

Cosmopolitan Species

This is the widest ranging of our phytogeographic elements and contains those taxa which have a worldwide distribution in both hemispheres. Only a small number of our taxa (27) belong to this element. Some of these include *Hutchinsia procumbens, Juncus bufonius, Potamogeton pectinatus, Scirpus cernuus, Triglochin maritima* and *Zostera marina.*

Bipolar Disjuncts

Plants in this element occur in both the Northern and Southern Hemispheres. This type of disjunction is not readily explained although Raven (1963) has suggested that most of these plants probably reached their disjunct areas by long-distance disperal relatively recently. A total of 41 BC taxa belong to this element. Typical examples include *Lilaea scilloides, Osmorhiza chilensis, Pectocarya penicillata, Sanicula graveolens, Scirpus olneyi* and *Trifolium microdon.*

Circumpolar Species

This group of widely distributed plants contains 16% (360 in total) of our flora. Although this element could be further subdivided into circumboreal, circumarctic and circumalpine elements (sensu Schofield 1969) we have not attempted that here. Representative species include *Carex pauciflora, Hierochloe odorata, Listera cordata, Potentilla hyparctica, Salix arctica, Saxifraga hieracifolia and Senecio congestus.*

Amphiberingian Species

This element includes species which occur on both sides of the Bering Sea, extending westward into Asia and eastward into North America. The role of the Bering Land Bridge in the distribution of northern plants has been the subject of numerous floristic papers (e.g., Hulten 1937, Porsild 1955, etc.). We have also included within this element a number of North Pacific Basin species that probably did not extend their ranges via the Bering Land Bridge but rather by the Aleutian Island Chain. Many of these North Pacific Basin species (i.e., *Carex macrochaeta, Mecodium wrightii* and *Plantathera chorisiana)* are readily identified but many others, due to more continental distribution are not. A total of 166 taxa (7% of the flora) belong to this amphiberingian element. Examples of these include *Arnica lessingii, Antennaria monocephala, Cardamine oligosperma, Carex membranacea, Rumex arcticus, Senecio tundricola* and *Spiraea stevenii.*

North American Maritime Species

Taxa belonging to this element occur along both the Pacific and Atlantic Coasts in North America and sometimes along the Arctic Coast. Some of the nine plants in this element include *Cakile edentula, Salicornia virginica, Senecio pseudoarnica, Spergularia canadensis* var. *canadensis* and *Puccinellia pumila.*

North American Radiants

Plants in this element range across large parts of North America. For this treatment, we have recognized this as a broad group, without subelements (i.e., American boreal, southern North American, etc.) Twenty-seven percent (a total of 606 taxa) occur in this element. Representative species include *Carex peckii, Plantanthera orbiculata, Primula mistassinica, Senecio pauperculus, Solidago multiradiata, Sporobolus asper* and *Vaccinium caespitosum.*

Cordilleran Species

This group contains species that occur primarily in the western Cordillera but which may have a small part of their range extending into more eastern or northern regions. These taxa comprise about 35% of the flora (a total of 794 taxa). Typical examples of Cordilleran species are *Arnica latifolia, Aster engelmannii, Draba albertina, Lupinus wyethi, Phacelia sericea, Rubus pedatus, Saxifraga mertensiana* and *Senecio cymbalarioides.*

Alaska–Yukon–Northwestern British Columbia Endemics

This group consists of species restricted to Alaska, Yukon and immediately adjacent northwestern British Columbia and occasionally the western Northwest Territories. Most of these endemics probably originated in a large glacial refugium in northern Alaska and the Yukon. Some, however, must have persisted on unglaciated nunataks or other areas farther south (Douglas *et al.* 1981). A small number of these (22 taxa) have reached the mountains of extreme northwestern British Columbia. Representative taxa include *Astragalus nutzotinensis, Douglasia gormanii, Montia bostockii, Oxytropis huddelsonii, O. scammaniana, Salix setchelliana* and *Senecio ogotorukensis.*

Pacific Coast Endemics

This element consists of mesophytic Pacific Coast species which have their ranges extending over large parts of the Pacific Coast between Alaska and Mexico. The 228 taxa (10% of the flora) in this Pacific Coast element may be further subdivided into three subelements:

Entire Pacific Coast Endemics

These taxa have ranges which extend from Alaska to Oregon, California or Mexico. This group consists of 54 taxa and includes species such as *Cardamine angulata, Carex obnupta, Malus fusca, Oenanthe sarmentosa, Oxalis oregana, Poa laxiflora, Ribes bracteosum* and *Sedum oreganum.*

North Pacific Coast Endemics

This subelement includes 30 plants which have ranges between Vancouver Island and Alaska. Ten of these plants have their main distribution in the Queen Charlotte Islands (Calder and Taylor 1968). Representative species include *Atriplex alaskanus, Castilleja unalaschcensis, Gentiana douglasiana, Lloydia serotina* ssp. *flava, Lupinus nootkatensis, Polystichum setigerum, Salix reticulata* ssp. *glabellicarpa* and *Senecio moresbiensis.*

South Pacific Coast Endemics

The South Pacific Coast subelement contains taxa with ranges that extend from Vancouver Island (usually the southern part of the island) southward varying distances to Mexico. In British Columbia this subelement consists of about five percent of the flora (144 taxa) and is represented by taxa such as *Artemisia suksdorfii, Aster chilensis, Lomatium utriculatum, Sanicula arctopoides, S. bipinnatifida, Senecio macounii* and *Trifolium depauperatum.* A few members of this subelement (*Bidens amplissima, Limnanthes macounii* and *Aster paucicapitatus)* do not have ranges extending beyond southwestern British Columbia or northwestern Washington.

COMPOSITION OF THE FLORA

The vascular plant flora of British Columbia comprises 132 families. Within these families are a total of 2854 taxa, 2301 of these are native while the remaining 553 are introduced (Table 1). Information on the distribution of the flora with respect to vegetation zonation is presented in Table 2.

TABLE 1. Families and number of genera, species, subspecies, varieties, taxa, and introduced taxa in the flora of British Columbia.

	Number of					
	Genera	**Species**	**Ssp.**	**Vars.**	**Taxa**	**Introduced Taxa**
Gymnosperms						
1. Cupressaceae	3	5	0	0	5	0
2. Pinaceae	6	19	0	4	21	0
3. Taxaceae	1	1	0	0	1	0
Total	**10**	**25**	**0**	**4**	**27**	**0**
Dicotyledons						
1. Aceraceae	1	6	0	0	6	2
2. Adoxaceae	1	1	0	0	1	0
3. Amaranthaceae	1	4	0	0	4	0
4. Anacardaceae	1	3	0	0	3	0
5. Apiaceae	29	59	4	7	62	12
6. Apocynaceae	2	6	0	4	7	2
7. Aquifoliaceae	1	1	0	0	1	1
8. Araliaceae	3	3	0	0	3	1
9. Aristolochiaceae	1	1	0	0	1	0
10. Asclepiadaceae	1	2	0	0	2	0
11. Asteraceae	81	310	48	77	351	84
12. Balsaminaceae	1	6	0	0	6	2
13. Berberidaceae	3	5	0	0	5	1
14. Betulaceae	3	12	2	4	15	3
15. Boraginaceae	15	44	3	3	45	14
16. Brassicaceae	48	139	4	28	152	49
17. Buddlejaceae	1	1	0	0	1	1
18. Cactaceae	1	2	0	0	2	0
19. Callitrichaceae	1	6	1	0	6	1
20. Campanulaceae	6	15	0	2	16	4
21. Cannabaceae	1	1	0	0	1	1
22. Capparidaceae	1	1	0	0	1	0
23. Caprifoliaceae	5	16	2	5	20	1
24. Caryophyllaceae	22	80	13	12	88	34
25. Celastraceae	2	2	0	0	2	0
26. Ceratophyllaceae	1	2	0	0	2	0
27. Chenopodiaceae	10	28	5	3	31	16
28. Convolvulaceae	2	8	0	0	8	5
29. Cornaceae	1	4	0	0	4	0
30. Crassulaceae	2	10	1	4	12	1
31. Cucurbitaceae	2	2	0	0	2	1

Table 1 *continued*

		Number of					
		Genera	Species	Ssp.	Vars.	Taxa	Introduced Taxa
32.	Diapensiaceae	1	1	0	0	1	1
33.	Dipsacaceae	3	3	0	0	3	3
34.	Droseraceae	1	2	0	0	2	0
35.	Eleagnaceae	2	4	0	0	4	1
36.	Elatinaceae	1	1	0	0	1	0
37.	Empetraceae	1	1	0	0	1	0
38.	Ericaceae	25	58	6	8	64	1
39.	Euphorbiaceae	1	9	0	0	9	7
40.	Fabaceae	19	125	14	20	137	38
41.	Fagaceae	1	2	0	0	2	1
42.	Fumariaceae	3	8	1	0	8	2
43.	Gentianaceae	6	17	5	0	17	1
44.	Geraniaceae	2	10	0	1	10	4
45.	Grossulariaceae	1	16	2	2	18	0
46.	Haloragaceae	1	9	0	0	9	3
47.	Hippuridaceae	1	3	0	0	3	0
48.	Hydrangeaceae	1	1	0	0	1	0
49.	Hydrophyllaceae	6	17	0	3	17	0
50.	Hyperiacaceae	1	4	0	2	5	1
51.	Lamiaceae	23	43	5	8	48	26
52.	Lentibulariaceae	2	6	0	2	7	0
53.	Limnanthaceae	2	2	0	0	2	0
54.	Linaceae	1	3	0	0	3	2
55.	Loasaceae	1	3	0	2	4	1
56.	Loranthaceae	1	3	0	4	2	0
57.	Lythraceae	4	5	0	0	5	2
58.	Malvaceae	4	10	0	1	10	5
59.	Menyanthaceae	2	2	0	0	2	0
60.	Molluginaceae	1	1	0	0	1	1
61.	Moraceae	1	1	0	0	1	1
62.	Myricaceae	1	2	0	0	2	0
63.	Nyctaginaceae	2	3	0	0	3	1
64.	Nymphaeaceae	3	7	0	0	7	3
65.	Oleaceae	2	2	0	0	2	1
66.	Onagraceae	9	42	7	2	47	4
67.	Orobanchaceae	2	7	1	3	8	0
68.	Oxalidaceae	1	3	0	0	3	2
69.	Papaveraceae	4	9	1	0	9	4
70.	Parnassiaceae	1	4	0	0	4	0
71.	Plantaginaceae	1	11	1	0	11	5
72.	Plumbaginaceae	1	1	2	0	2	0
73.	Polemoniaceae	9	26	3	2	27	0
73.	Polygalaceae	1	1	0	0	1	0
74.	Polygonaceae	6	47	11	8	55	14
75.	Portulaceae	7	25	0	6	26	1
76.	Primulaceae	10	34	4	4	37	4

Table 1 *continued*

		Number of					
		Genera	Species	Ssp.	Vars.	Taxa	Introduced Taxa
77.	Ranunculaceae	16	80	8	17	86	8
78.	Resedaceae	1	2	0	0	2	2
79.	Rhamnaceae	2	4	0	2	5	0
80.	Rosaceae	27	109	27	28	136	23
81.	Rubaceae	2	12	3	0	14	3
82.	Salicaceae	2	47	14	7	57	3
83.	Santalaceae	2	2	0	2	3	0
84.	Sarraceniaceae	1	1	0	0	1	0
85.	Saxifragaceae	12	51	7	10	58	1
86.	Scrophlariaceae	22	109	7	20	116	20
87.	Solanaceae	3	7	0	1	7	6
88.	Thymelaeaceae	1	1	0	0	1	1
89.	Ulmaceae	1	1	0	0	1	1
90.	Urticaceae	2	3	1	0	3	1
91.	Valerianaceae	3	9	3	0	9	2
92.	Verbenaceae	1	2	0	0	2	2
93.	Violaceae	1	22	5	6	24	3
94.	Zygophyllaceae	1	0	0	0	0	0
Total		**529**	**1836**	**221**	**320**	**2026**	**452**
Pteridophytes							
1.	Adiantaceae	7	12	0	2	13	0
2.	Aspleniaceae	1	3	0	0	3	0
3.	Blechnaceae	2	2	0	0	2	0
4.	Dennstaedtiaceae	1	1	2	0	2	0
5.	Dryopteridaceae	7	28	2	0	28	0
6.	Hymenophyllaceae	1	1	0	0	1	0
7.	Polypodiaceae	1	5	0	0	5	0
8.	Thelypteridaceae	2	3	0	0	3	0
9.	Marsileaceae	1	1	0	0	1	0
10.	Azollaceae	1	2	0	0	2	1
11.	Lycopodiaceae	3	10	0	3	12	0
12.	Selaginellaceae	1	5	0	2	6	0
13.	Equisetaceae	1	10	1	2	11	0
14.	Ophlioglossaceae	2	14	0	0	14	0
15.	Isoetaceae	1	7	0	0	7	0
Total		**32**	**104**	**5**	**9**	**110**	**1**

Table 1 *continued*

	Number of					
	Genera	Species	Ssp.	Vars.	Taxa	Introduced Taxa
Monocotyledons						
1. Alismataceae	2	4	0	3	5	0
2. Araceae	3	3	0	0	3	0
3. Butomaceae	1	1	0	0	1	1
4. Cyperaceae	10	185	18	12	198	2
5. Hydrocharitaceae	3	4	0	0	4	2
6. Iridaceae	4	10	0	4	12	4
7. Juncaceae	2	45	8	8	54	0
8. Juncaginaceae	1	3	0	1	3	0
9. Lemnaceae	3	5	0	0	5	0
10. Lilaeaceae	1	1	0	0	1	0
11. Liliaceae	24	53	1	17	56	5
12. Najadaceae	1	1	0	0	1	0
13. Orchidaceae	15	31	4	6	35	1
14. Poaceae	71	244	40	17	271	82
15. Pontederiaceae	1	1	0	0	1	0
16. Potamogetonaceae	1	23	0	0	23	1
17. Ruppiaceae	1	1	0	2	2	0
18. Scheuchzeriaceae	1	1	0	1	1	0
19. Sparganiaceae	1	7	1	0	7	0
20. Typhaceae	1	2	0	0	2	1
21. Zannichelliaceae	1	1	0	0	1	0
22. Zosteraceae	2	5	0	0	5	1
Total	**149**	**631**	**72**	**71**	**691**	**98**
Grand Total	**720**	**2595**	**298**	**404**	**2854**	**553**

TABLE 2. Distribution of vascular plants with respect to vegetation zones in British Columbia.

Vegetation Zone	Number of Taxa Restricted to Zone	Number of Taxa Occurring in Zone
Alpine	102	588
Subalpine	10	733
Steppe	137	1190
Montane	232	1827
Lowland	478	1480

ALISMATACEAE

1. Achenes borne in a single whorl on the receptacles; stamens 6 (rarely 9) .... *Alisma*

1. Achenes densely packed over entire surface of the receptacles; stamens 7–25 .... *Sagittaria*

ALISMA

1. Achenes centrally grooved toward the tip; petals 3.5–6 mm long, white .... *A. plantago-aquatica*

1. Achenes centrally ridged and with two groves towards the tip; petals 2–4 mm long, pinkish or whitish .... *A. gramineum*

***Alisma gramineum* J.G. Gmel.**

Narrow-leaved water-plaintain

Habitat/Range: Lakeshores, rivers, marshes and tidal flats in the lowland, steppe and montane zones; infrequent in S BC; circumpolar, E to S PQ and S to NY, NM, AZ and N CA; Eurasia.

Notes: The taxonomy of this species is extremely complex. At present we are recognizing two varieties in BC.[1] For a discussion of the nomenclatural problems in this species see Reveal (1977).

1. Leaves largely submersed or floating, linear, almost without a blade, but plants sometimes emergent and with narrowly elliptic blades rarely as much as 5 mm wide; plants occurring E of the Coast-Cascade Mountains in BC .... var. *gramineum*

1. Leaves strictly erect, rarely submersed, blades mostly more than 5 mm wide; plants occurring W of the Coast-Cascade Mountains in BC .... var. *angustissimum* Hendricks

***Alisma plantago-aquatica* L. var. *americana* Schultes & Schultes** (*A. plantago-aquatica* var. *triviale* [Britt., Sterns & Pogg.] Far.)

American water-plantain

Habitat/Range: Shallow ponds, marshes and ditches in the lowland, steppe and montane zones; common in S BC south of 52°C; rare northward; E to NF and S to most of the US.

Notes: Despite several monographs on *Alisma* the taxonomy and nomenclature of the above species still remains confused (see Reveal [1977]).

SAGITTARIA

1. Achenes with beaks extending at right angles to the body, the beaks 0.5–1.5 mm long; bracts of the inflorescences ovate, rounded or hooded, 5–10 (15) mm long; mature staminate and pistillate pedicels of equal length .... *S. latifolia*

1. Achenes with beaks erect, the beaks 0.2–0.4 mm long; bracts of the inflorescences lanceolate, acuminate (5) 10–30 mm long; mature staminate pedicels longer than the pistillate pedicels .... *S. cuneata*

***Sagittaria cuneata* Sheldon** (*S. arifolia* Nutt.)

Arum-leaved arrowhead

Habitat/Range: Wet ditches, ponds and lakes in the lowland, steppe and montane zones; common in BC south of 55°N and E of the Coast-Cascade Mountains, less frequent westward and northward; N to AK, YT and NT, E to Labr., NB and NS and S to most of US.

---

1 Key adapted from Hitchcock (1969).

Sagittaria latifolia* Willd. var. *latifolia (*S. esculenta* Howell)
Wapato
Habitat/Range: Wet ditches, ponds, lakeshores and marshes in the lowland, steppe and montane zones; infrequent in BC south of 56°N, absent from Queen Charlotte Islands, N Vancouver Island and adjacent coast; E to PQ, NB and NS and S to most of the US; N South America, HI.

ARACEAE

1. Leaves linear, less than 3 cm wide; spathes leaflike, green, not enclosing the spadix .... *Acorus*
1. Leaves lanceolate to ovate, more than 3 cm wide; spathes not leaflike, yellow or white, enclosing the spadix.
   2. Leaves net-veined, lanceolate to oblong-ovate; spathes yellow .... *Lysichiton*
   2. Leaves pinnately closely parallel-veined, heart-shaped; spathes white .... *Calla*

ACORUS

***Acorus americanus* (Raf.) Raf.** (*A. calamus* auct. non L.)
American sweet-flag
Habitat/Range: Shallow water in the montane zone; rare in S BC east of the Coast-Cascade Mountains; N to NT, E to NF and S to WA, ID, MT and E US.
Notes: Packer and Ringius (1984) have shown that our fertile diploid species is different than the sterile triploid Eurasian species (*A. calamus* L.) which is introduced in E North America.

CALLA

***Calla palustris* L.**
Wild calla
Habitat/Range: Swamps, marshes and lakeshores in the montane zone; frequent in BC north of 52°N and E of the Coast Mountains, rare in SC BC; circumpolar, N to AK, YT and NT, E to NF and S to TX, FL and CO; Eurasia.

LYSICHITON

***Lysichiton americanum* Hult. & St. John**
Skunk cabbage
Habitat/Range: Swamps, wet ditches and moist forests in the lowland and montane zones; common W of the Coast-Cascade Mountains, less frequent in the Skeena, upper Fraser and Columbia River drainages; N to SE AK and S to MT, ID and CA.

BUTOMACEAE

BUTOMUS

***Butomus umbellatus* L.**
Flowering-rush
Habitat/Range: Marshes and lakeshores; rare in SW BC, known only from Hatzic Lake; introduced from Eurasia.

CYPERACEAE

1. Flowers unisexual; achenes enclosed or enwrapped in a perigynium.
   2. Perigynium open, with unsealed margins, merely wrapped around the achene .... *Kobresia*
   2. Perigynium closed, except for the apex .... *Carex*
1. Flowers perfect; achenes not enclosed or enwrapped in a perigynium.
   3. Scales of spikelets in 2 vertical rows.
      4. Perianth of 6–9 bristles; leaves numerous, equally distributed on the stems, upper ones bearing axillary inflorescences .... *Dulichium*
      4. Perianth bristles absent; leaves fewer, more basally disposed; inflorescences terminal .... *Cyperus*
   3. Scales of spikelets spirally arranged.
      5. Spikelets with 1 (rarely 2) achenes .... *Rhynchospora*
      5. Spikelets with several to many achenes.
         6. Styles thickened toward the base, this thickening forming a persistent and conspicuous tubercle on the achenes .... *Eleocharis*
         6. Styles not thickened although achenes may be apiculate.
            7. Achenes subtended by inconspicuous, adaxial scales in addition to the ordinary scales of the spikelets .... *Hemicarpha*
            7. Achenes subtended by 1–many bristles (or these rarely obsolete) in addition to the ordinary scales of the spikelets.
               8. Perianth bristles 10 or more .... *Eriophorum*
               8. Perianth bristles 8 or fewer.
                  9. Spikelets with subtending, leafy bracts .... *Scirpus*
                  9. Spikelets with slightly modified lower scales, not leafy.
                     10. Lower scales of the spikelets short, not more than half as long as the spikelet .... *Eleocharis*
                     10. Lower scales of the spikelets long, nearly as long or longer than the spikelet .... *Trichophorum*

CAREX[2]

1. Spike solitary, terminal .... GROUP A
1. Spikes more than one, sometimes congested into a head-like inflorescence.
   2. Spikes all alike, usually having both staminate and pistillate flowers or plants dioecious; all spikes sessile .... GROUP B
   2. Spikes differentiated into terminal staminate spike and lateral pistillate spikes; if terminal spike bisexual, then lateral spikes have noticeable peduncles .... GROUP C

---

2 Contributed by George W. Douglas and A. Ceska; key by A. Ceska.

GROUP A

Sedges with a single terminal spike

1. Plants dioecious, spikes with either male or female flowers.
   2. Perigynia glabrous, at maturity reflexed; stigmas two; leaves involute, less than 1 mm wide .... *C. gynocrates*
   2. Perigynia hairy, at maturity not reflexed; stigmas three; leaves flat, at least 1 mm wide .... *C. scirpoidea*
1. Plants monoecious, spikes with both male and female flowers.
   3. Lower pistillate scales leaflike, much longer than the perigynia and even longer than spike .... *C. saximontanus*
   3. Lower pistillate scales not leaflike, at most slightly longer than the perigynia.
      4. Mature perigynia reflexed, scales deciduous.
         5. Perigynia ovate or broadly lanceolate in outline, 3–4 mm long.
            6. Leaves flat, 1.5–2.5 mm wide; stigmas 3 .... *C. nigricans*
            6. Leaves cylindrical or involute, less than 1 mm wide; stigmas 2 .... *C. gynocrates*
         5. Perigynia narrowly lanceolate in outline, 4–7 mm long.
            7. Perigynia 6–7 mm long, without nerves; rachilla not exserted from the perigynium; culms trigonous, scabrous, aphyllopodic .... *C. pauciflora*
            7. Perigynia 4–5.5 mm long, with few nerves; rachilla conspicuously exserted from the perigynium; culms rounded, smooth, phyllopodic .... *C. microglochin*
      4. Mature perigynia not reflexed, scales persistent.
         8. Perigynia rounded at apex, beakless .... *C. leptalea*
         8. Perigynia not rounded at the apex, contracted or tapered into a beak.
            9. Plants rhizomatous.
               10. Perigynia oblong or lanceolate, 4 mm long or longer.
                   11. Perigynia 4 mm long, nerved; rhizomes light brown-green .... *C. anthoxanthea*
                   11. Perigynia 5–7 mm long, 2-ribbed; rhizomes reddish-brown .... *C. geyeri*
               10. Perigynia round to obovate, up to 4 mm long.
                   12. Pistillate scales about 1/2 as long as perigynia, acute to short-awned; perigynia coriaceous; rachilla well developed .... *C. obtusata*
                   12. Pistillate scales about as long as perigynia, obtuse; perigynia membranaceous; rachilla absent .... *C. rupestris*
            9. Plants tufted.
               13. Perigynia pubescent or puberulent, at least on the beaks .... *C. filifolia*
               13. Perigynia glabrous.
                   14. Spikes rounded at the base; perigynia ovate in outline, rounded at the base, without a short stipe.

15. Spikes globose, 4–10 mm long; stigmas 2; perigynia not conspicuously inflated, abruptly contracted into a short beak .... *C. capitata*

15. Spikes elongated, 10–15 mm long, stigmas 3; perigynia strongly inflated, continuously tapering into a minute beak ... *C. engelmannii*

14. Spikes tapering towards the base; perigynia lanceolate or oblanceolate in outline with a short stipe.

16. Plants densely tufted with conspicuous remnants of old sheaths without blades; spikes ovate, shorter than 10 mm; rachilla present .. .. *C. nardina*

16. Plants loosely tufted without remnants of old sheaths; spikes elliptical to linear, usually longer than 10 mm; rachilla absent.

17. Staminate part of the spikes consisting of several flowers and forming less than 1/5 of the spike; perigynia and scales dark tinged; beaks of the perigynia glabrous on sides; spikes dense, ovate to lanceolate, 2–4 times longer than wide .... *C. pyrenaica*

17. Staminate part of the spikes conspicuous, forming 1/5–1/3 of the spike; perigynia and scales light brown; beaks scabrous with scattered spines on the sides; spikes narrowly lanceolate, more than 4 times longer than wide .... *C. circinata*

GROUP B

Sedges with several spikes.

Each spike with both male and female flowers.

1. Perigynia large, 10–15 mm long, with 5 mm long beaks; stigmas 3 .... *C. macrocephala*

1. Perigynia smaller than 7 mm, beakless or with beaks shorter than 3 mm; stigmas 2.

2. Male flowers at the upper part of each individual spike (in mature plants look for remnants of filaments or for empty scales) or plants dioecious.

3. Culms arising singly or few together from long creeping rhizomes.

4. Culms branched, becoming decumbent and sending up new culms from the axils of old leaves .... *C. chordorrhiza*

4. Culms unbranched.

5. Perigynia wing-margined; beak acutely bidentate .... *C. foenea*

5. Perigynia not wing-margined; beaks obliquely cut or sometimes bidentate.

6. Spikes densely aggregated into a globose or ovoid head appearing like one spike .... *C. maritima*

6. Spikes not densely aggregated, at least the lower ones distinct.

7. Spikes few-flowered, with 2–3 perigynia; perigynia with inconspicuous beaks .... *C. disperma*

7. Spikes with more than 3 perigynia per spike; perigynia with evident beak.

8. Upper sheaths green-striate ventrally; perigynia with beaks sharply bidentate, thin-margined above .... *C. sartwellii*

8. Upper sheaths hyaline ventrally; perigynia with beaks not sharply bidentate; perigynia not thin-margined.

 9. Culms obtusely angled or obtusely triangular, smooth; leaves narrowly involute or canaliculate.

  10. Perigynia long-beaked; styles elongate .... *C. douglasii*

  10. Perigynia short beaked; styles short .... *C. eleocharis*

 9. Culms sharply triangular, rough above; leaves flat or channelled.

  11. Perigynia chestnut colored, unequally biconvex; beak 1/5–1/3 of the body; rootstock slender; lower sheaths light brownish .... *C. simulata*

  11. Perigynia blackish in age, plano-convex; beak 1/3–1/2 of the body; rootstock stout; lower sheaths dark brown or black.

   12. Scales very dark, chestnut brown, shining; perigynia 3.5–4.5 mm long, polished, scarcely hyaline at orifice .... *C. pansa*

   12. Scales lighter colored, dull; perigynia 3–4 mm, dull, strongly hyaline at orifice .... *C. praegracilis*

3. Culms densely to loosely cespitose, rootstock not long-creeping.

13. Perigynia with body tapering into the beak; culms weakly cellular and flattened in drying; leaves 5–10 mm wide .... *C. stipata*

13. Perigynia with body abruptly contracted into the beak; culms not weakly cellular and not flattening in drying; leaves 1–5 mm wide.

 14. Spikes usually fewer than ten, congested into a subglobose or broadly cylindrical head (occasionally with one or two remote spikes); heads 1–3 times longer than wide; sheaths not red-dotted ventrally .... *C. hoodii*

 14. Spikes more than ten, aggregated in interrupted heads; heads more than 4 times longer than wide; sheaths often red-dotted ventrally, at least at mouth.

  15. Perigynia yellowish or straw-colored, or in age black, plano-convex; leaf-sheaths usually transversely wrinkled ventrally; the lowermost spikes with conspicuous bristle-like bracts, much longer than the corresponding spikes .. .. *C. vulpinoidea*

  15. Perigynia brownish, thick plano-convex or unequally biconvex; leaf-sheaths not transversely wrinkled ventrally; bracts absent or at most as long as the corresponding spikes.

   16. Sheaths not copper-tinged at the mouth; heads little interrupted; perigynia 2.0–2.75 mm long, shining, not concealed by the scales .... *C. diandra*

   16. Sheaths copper-tinged at the mouth; heads interrupted; perigynia 2.5–4.0 mm long, nearly concealed by the scales.

17. Inflorescences branched; leaves 2.5–6.0 mm wide, ligules conspicuous, longer than wide; sheaths concave at mouth and only moderately prolonged beyond the base of blades; perigynia 3–4 mm long, shiny; beak setulose-serrulate .... *C. cusickii*

17. Inflorescences not branched; leaves 2–3 mm wide, ligules very short; sheaths convex and conspicuously prolonged beyond the base of the blades; perigynia 2.5–3.0 mm long, dull; beak serrulate .... *C. prairea*

2. Male flowers at the base of each individual spike (in mature plants look for remnants of filaments or for empty scales).

18. Perigynia narrowly to broadly wing-margined, the lower part of the body not spongy thickened.

19. Bracts leaf-like, conspicuously exceeding the head.

20. Perigynia very narrow, 4.8–6.5 mm long and 0.8–1.0 mm wide, 5–7 times as long as wide .... *C. sychnocephala*

20. Perigynia 3.0–4.5 mm long and 0.9–1.8 mm wide, 2–4.5 times as long as wide.

21. Perigynia with beak hyaline at orifice, obliquely cleft, shallowly bidentate; lowest bract not appearing like a continuation of the culm .... *C. athrostachya*

21. Perigynia with beak not hyaline at orifice, bidentate; lowest bract appearing like a continuation of the culm .... *C. unilateralis*

19. Bracts not leaflike nor conspicuously exceeding head.

22. Perigynia with beak slender and nearly terete and scarcely margined at the tip, the upper 1–2 mm little if at all serrulate, obliquely cut dorsally, at times becoming bidentulate or rarely bidentate.

23. Perigynia 6–8 mm long; scales light reddish-brown; spikes more or less aggregated into an erect head .... *C. petasata*

23. Perigynia 2.5–6.0 mm long, if longer, the scales very dark-colored or spikes in a flexuous or moniliform inflorescence.

24. Scales shorter than perigynia and noticeably narrower above, largely exposing perigynia above.

25. Achenes filling more than one 1/2 of the perigynia.

26. Perigynia 2.5–3.5 mm long; spikes closely aggregated into a small black head .... *C. illota*

26. Perigynia 3.5–6 mm long; spikes loosely aggregated into a green to dark brown head.

27. Perigynia plump, biconvex; the achenes filling about 3/4 the perigynium body; anthers about 1.5 mm long ... . *C. preslii*

27. Perigynia planoconvex; achene filling 1/2 to 2/3 the perigynium; anthers 0.7–1.0 mm long.

28. Scales with a broad hyaline margin; perigynium wings copper-coloured, darker than the perigynium bodies .... *C. macloviana*

28. Scales with a narrow hyaline margin or dark throughout; perigynium wings the same colour as the rest of the perigynium bodies .... *C. pachystachya*

25. Achenes filling less than 1/2 the perigynia.

29. Perigynia 4.5–6.5 mm long; spikes forming heads triangular in the outline; heads black .... *C. haydeniana*

29. Perigynia 3–4.3 mm long; spikes forming heads ovate in outline; heads brown .... *C. microptera*

24. Scales about the length of perigynia and nearly the same width above, concealing perigynia above or nearly so.

30. Perigynia with beak not white-hyaline at apex .... *C. tracyi*

30. Perigynia with beak strongly white-hyaline at apex.

31. Culms 1–3 dm high; pistillate scales reddish-brown without the hyaline margin; spikes 4–6 mm long ... *C. phaeocephala*

31. Culms 2–8 dm high; pistillate scales light brown with broad, silvery-white hyaline margins; spikes usually 6–16 mm long .... *C. praticola*

22. Perigynia with beak flattened and margined at tip, serrulate to the apex, often strongly bidentate as well as obliquely cut dorsally.

32. Scales about as long as the perigynia and nearly the same width above, concealing perigynia above or very nearly so.

33. Inflorescences slender, flexuous or moniliform; scales light brown or brown with a broad hyaline margin .... *C. aenea*

33. Inflorescences stiff, spikes approximate or aggregated; scales silvery throughout .... *C. xerantica*

32. Scales shorter than perigynia and noticeably narrower above, largely exposing perigynia above.

34. Leaf-sheaths (at least of uppermost leaves) green and strongly nerved ventrally nearly or quite to the mouth .... *C. feta*

34. Leaf-sheaths strongly white-hyaline ventrally.

35. Perigynia subulate to narrowly ovate-lanceolate, 3–4 times as long as wide.

36. Perigynia 3–4 mm long, lanceolate-subulate, where distended by the achene 0.8–1.2 mm wide, their margins nearly obsolete at the base .... *C. crawfordii*

36. Perigynia 4–7 mm long, lanceolate to narrowly ovate-lanceolate, 1.2–2.6 mm wide, noticeably wing-margined to the base .... *C. scoparia*

35. Perigynia ovate-lanceolate or broader, at most twice as long as wide.

37. Perigynia 3–5 mm long, 2.0–3.5 mm wide, the body suborbicular to reniform .... *C. brevior*

37. Perigynia 3.0–4.5 mm long, 1.5–2.0 mm wide, the body narrowly to broadly ovate.

38. Perigynia 3.0–3.5 mm long, brownish at maturity (dull green when young); spikes closely aggregated, truncate or rounded at base .... *C. bebbii*

38. Perigynia 2.8–4.0 mm long, straw-coloured at maturity (green when young); at least the lowest two spikes separated, spikes short clavate at base .... *C. tenera*

18. Perigynia at most thin-edged, the lower part of the body spongy thickened.

39. Perigynia not whitish-puncticulate.

40. Perigynia appressed; perigynium walls thin, papery, breaking easily and exposing the achene .... *C. deweyana*

40. Perigynia spreading or ascending at maturity; perigynium walls thicker, not breaking easily and exposing the achene.

41. Beak smooth or nearly so, body broadest near the middle; spikes remote, the lowest ones with bracts much longer than spikes .... *C. laeviculmis*

41. Beak serrulate, body broadest near the base; spikes close together in the upper part of the culm; bract short or none.

42. Perigynia small, 2.25–3.25 mm long; beaks very shallowly bidentate, 1/4 to 1/3 length of the body; scales 1/2 to 2/3 length of the body .... *C. interior*

42. Perigynia 2.75–4.75 mm long; beaks sharply bidentate and ventral false suture conspicuous .... *C. echinata*

39. Perigynia whitish-puncticulate.

43. Plants with stolons and rhizomes forming loose mats or growing as single culms.

44. Spikes with fewer than 4 perigynia, widely separated on the culm.

45. Perigynia unequally biconvex; spikes essentially androgynous; lowest bract short or none .... *C. disperma*

45. Perigynia planoconvex; spikes gynaecandrous; lowest bract bristle-form, many times exceeding its spike .... *C. trisperma*

44. Spikes 4–10 flowered, lower spikes sometimes remote, the upper close together, or in a terminal head.

46. Spikes closely aggregated, forming an ovoid or suborbicular head; perigynia with slender nerves .... *C. tenuiflora*

46. Spikes remote; perigynia many-ribbed .... *C. loliacea*

43. Plants cespitose, tufted.

47. Perigynia ovate, broadest near the base; with beak conspicuous, about 0.7 mm long, strongly serrulate .... *C. arcta*

47. Perigynia elliptic, broadest near the middle; beak shorter than 0.5 mm, smooth or moderately serrulate.

48. Perigynia beakless or with a short blunt beak without a hyaline-margined suture.

49. Lateral spikes pistillate without staminate flowers; terminal spike gynaecandrous and longer and larger than the lateral spikes .... *C. glareosa*

49. Lateral spikes gynaecandrous, similar in shape and size to the terminal spikes.

50. Perigynia 2.3–3 mm long with 6–10 fine nerves on each side .... *C. canescens*

50. Perigynia 1.5–2 mm long with 5–7 fine nerves on each side .... *C. lapponica*

48. Perigynia with a conspicuous sharp beak with a distinct hyaline-margined suture dorsally.

51. Spikes 3–9, at least the lower ones remote .... *C. brunnescens*

51. Spikes 1–4(6), closely approximate.

52. Culms smooth or nearly so below the head; perigynia brownish-yellow; the beak hyaline at apex ...... *C. lachenallii*

52. Culms very rough below the head; perigynia cinereous or brown in age; the beak reddish brown at apex .... *C. heleonastes*

GROUP C

Sedges with terminal staminate and lateral pistillate spikes

1. Stigmas two; achenes lenticular.

2. Lowest bract sheathing; spike few-flowered; perigynia pulverulent or golden-yellow at maturity.

3. Lowest bract slightly exceeding culm, short-sheathing; terminal spikes gynaecandrous with at least 3/4 of flowers female .... *C. bicolor*

3. Lowest bract strongly exceeding culm, long-sheathing; terminal spikes male or gynaecandrous with fewer than 3/4 of flowers female.

4. Mature perigynia golden yellow or brownish, coarsely ribbed; scales widely spreading ... . *C. aurea*

4. Mature perigynia whitish pulverulent, obscurely ribbed; scales appressed ...... *C. garberi*

2. Lowest bract usually sheathless; spikes many-flowered; perigynia not pulverulent or golden yellow at maturity.

5. Perigynia lustrous, over 4 mm long; achenes continuous with style .... *C. saxatilis*

5. Perigynia not lustrous, smaller than 4 mm; achenes jointed with style.

6. Perigynia conspicuously nerved or ribbed .... *C. lenticularis*

6. Perigynia nerveless ventrally or with obscure impressed nerves.

7. Perigynia leathery (coriaceous); achenes constricted in the middle.

8. Pistillate spikes erect on stiff peduncles; perigynia plano-convex, nerveless; sheaths filamentous .... *C. obnupta*

8. Pistillate spikes pendulous on slender elongate peduncles; perigynia unequally bi-convex, coriaceous, nerveless on dorsal side and with faint nerves on ventral side; sheaths not filamentous .... *C. lyngbyei*

7. Perigynia papery (chartaceous); achenes not constricted in the middle.

9. Perigynia ovate-orbicular, olive-green, 1.5–2 mm long .... *C. interrupta*

9. Perigynia narrowly elliptic to broadly ovate or obovate, light green to straw-colored, 2.5–3.5 mm long.

10. Lowest bract shorter than inflorescence; pistillate scales with obsolete or slender midvein.

11. Pistillate spikes slender, linear to narrowly oblong, 2–4 (6) mm wide, the lowest pedunculate; perigynia plano-convex, appressed .... *C. bigelowii*

11. Pistillate spikes stout, oblong, 6–10 mm wide, densely flowered, subsessile; perigynia soon turgid, spreading .... *C. scopulorum*

10. Lowest bract equaling or exceeding inflorescence; pistillate scales with slender midvein or broader light-coloured center.

12. Perigynia inflated; spikes truncate at the base .... *C. aperta*

12. Perigynia flattened; spikes cuneate at the base.

13. Sheaths usually strongly dark-colored ventrally at mouth; lower pistillate spikes nodding on long peduncles; pistillate scales whitened at tip, especially in age .... *C. sitchensis*

13. Sheaths not colored ventrally at mouth; lower pistillate spikes not nodding; pistillate scales not whitened at tip ..... *C. aquatilis*

1. Stigmas three, achenes triangular.

14. Perigynia pubescent.

15. Beaks of perigynia bidentate.

16. Lowest bract with a long sheath.

17. Terminal spike usually androgynous (staminate above pistillate), the upper lateral spike sessile and staminate, androgynous or pistillate, the lower 1 or 2 spikes pistillate or androgynous; plants with slender horizontal stolons; perigynia ciliate.

18. Perigynia lanceolate, 1.5–1.75 mm wide, tapering to apex; spikes 2–5, the terminal one with but a few perigynia at base .... *C. petricosa*

18. Perigynia oblong-ovate, 2.0–2.5 mm wide, abruptly minutely beaked; spikes more numerous, the terminal one more strongly pistillate at base .... *C. franklinii*

17. Terminal spike staminate or gynaecandrous (pistillate above staminate), the lateral pistillate; plants densely cespitose or with short ascending stolons; perigynia ciliate-serrulate only on the beak margin.

19. Perigynia narrowly lanceolate, 1 mm wide, flattened-triangular, the upper half very long-tapering .... *C. misandra*

19. Perigynia narrowly lanceolate to ovate, 1.25–2.25 mm wide, contracted into a beak .... *C. ablata*

16. Lowest bract sheathless.

20. Pistillate spikes more than 10 mm long, with more than 25 perigynia; plants usually taller than 50 cm.

21. Leaves involute, not more than 2 mm wide, smooth; culms obtusely triangular .... *C. lasiocarpa*

21. Leaves flat, 2–5 mm wide, scabrous on margins; culms sharply triangular .. .. *C. lanuginosa*

20. Pistillate spikes less than 10 mm long, with fewer than 15 perigynia; plants less than 50 cm tall.

22. Lower pistillate spikes arising from the base of the culms on elongate weak peduncles, widely separated from the terminal spikes.

23. Bracts of lowest non-basal pistillate spikes leaflet-like, normally exceeding culm and, if coloured, purplish brown-tinged at base.

24. Rootstocks slender; culms normally loosely cespitose; perigynia 2.5–3.0 mm long, short beaked; shallowly bidentate; staminate spike 2.5–3.0 mm long, inconspicuous .... *C. deflexa*

24. Rootstocks stout; culms more or less densely cespitose; perigynia (2.5) 3.0–4.5 mm long, beaks usually more than 0.5 mm long; staminate spikes more than 5 mm long .... *C. rossii*

23. Bracts of lowest non-basal pistillate spikes squamiform and shorter than the culm or, if longer, auriculate and reddish brown-tinged at base.

25. Perigynia nearly orbicular in cross-section, 1.75–2.25 mm wide; beaks not strongly two-edged .... *C. brevicaulis*

25. Perigynia triangular-orbicular in cross-section, 1.0–1.5 mm wide; beaks strongly two-edged .... *C. rugosperma*

22. Lower pistillate spikes at most moderately separate from the upper ones, all developing on the upper part of the culm.

26. Perigynia up to 1.5 mm wide, oblong-ovoid or elliptical (excluding the stipitate base and beak); plants with short, intertwined rhizomes .... *C. peckii*

26. Perigynia 2.0–2.3 mm wide, suborbicular, orbicular-obovoid (excluding the base and the beak); plants with long well developed rhizomes .... *C. inops*

15. Perigynia beakless or with the beak obliquely cut.

27. Culms with one or two pistillate spikes from long, filiform basal peduncles; pistillate scales abruptly cuspidate or short-awned .... *C. pedunculata*

27. Culms without pistillate spikes at the base; pistillate scales not abruptly cuspidate.

28. Staminate spikes 3–6 mm long; pistillate scales obtuse, ciliate, one half as long as the perigynia; stigmas very short .... *C. concinna*

28. Staminate spikes 8–25 mm long; pistillate scales acute to acuminate, from somewhat shorter than to exceeding perigynia; stigmas long, slender.

29. Perigynia loosely pubescent, wider and longer than the ciliate scales; staminate spikes sessile or nearly so; pistillate spikes 5–10 flowered; bracts very short-sheathing; culms smooth or nearly so; stigmas usually four .... *C. concinnoides*

29. Perigynia appressed-pubescent, narrower and shorter than the smooth scales; staminate spikes noticeably pedunculate; pistillate spikes 10–25 flowered; lowest bracts long-sheathing, spatulate; culms strongly roughened above; stigmas three .... *C. richardsonii*

14. Perigynia glabrous.

30. Perigynia beakless or beaks not toothed. (See page 25 for the opposite lead!)

31. Lower bract long sheathing.

32. Lowest bract without blade; scales yellowish-white .... *C. eburnea*

32. Lowest bract with setaceous or leafy blade; scales green or brownish to purplish black.

33. Pistillate spikes on short peduncles or sessile, erect, globose to cylindrical.

34. Leaves narrow, 0.2–1.0 mm broad, canaliculate or involute; pistillate spike usually single, globose, with 1–5 perigynia .... *C. glacialis*

34. Leaves 2–15 mm wide, V-shaped or flat; pistillate spikes usually 2–7, elongated, with 5–20 perigynia.

35. Perigynia rounded at base, suborbicular in cross-section; achenes loosely enveloped .... *C. crawei*

35. Perigynia tapering at base, triangular; achenes closely enveloped.

36. Plants rhizomatous, bluish-green; leaves 2–3.5 mm wide, V-folded .... *C. livida*

36. Plants cespitose, green; leaves 6–15 mm wide, M-folded ... . *C. hendersonii*

33. Pistillate spikes on peduncles about as long as spikes or longer, elongate, linear to cylindrical.

37. Pistillate spikes dense and broad, more than 5 mm in diameter; perigynia 5–6 mm long, with beak as long as or longer than the perigynium body .... *C. sprengelii*

37. Pistillate spikes loose and slender, less than 5 mm in diameter; perigynia 2.5–5 mm long, with beak shorter than one half of the perigynium body.

38. Plants rhizomatous; perigynia 3.5–5 mm long; pistillate spikes loosely upright, not drooping .... *C. vaginata*

38. Plants cespitose; perigynia 2–3 mm long; at least the lowest spikes drooping.

39. Terminal spike gynaecandrous (pistillate above staminate) .... *C. krausei*

39. Terminal spike staminate .... *C. capillaris*

31. Lower bract sheathless or very short sheathing.

40. Roots with yellowish woolly hairs.

41. Perigynia suborbicular-triangular in cross-section, blackish when ripe .... *C. pluriflora*

41. Perigynia compressed-triangular in cross-section, green or blue-green when ripe.

42. Pistillate scales terminating in a slender rough awn, 2–12 mm long .... *C. macrochaeta*

42. Pistillate scales attenuate, short cuspidate, or obtuse, but not with a long awn.

43. Staminate spikes 15–27 mm long; pistillate scales obtuse or acute, as broad as or broader than the perigynia and barely exceeding them in length, persistent; leaves glaucous, involute; plants with prolonged, slender rhizomes .... *C. limosa*

43. Staminate spikes 4–12 mm long; pistillate scales long attenuate, much narrower and longer than the perigynia, early deciduous; leaves scarcely glaucous, not involute; plants densely cespitose .... *C. magellanica*

40. Roots without yellowish hair cover.

44. Terminal spike staminate.

45. Perigynia nearly round or obtusely triangular in cross-section, little or not at all flattened.

46. Plants shorter than 20 cm, rhizomatous; bracts of the lowest spike with conspicuous hyaline auricles .... *C. supina*

46. Plants taller than 20 cm, cespitose; bracts of the lowest spike (if developed at all) without auricles.

47. Perigynia beakless, lustrous, green; lower leaf sheaths puberulent .... *C. pallescens*

47. Perigynia with a short beak, puncticulate, blue-green or reddish-brown, firm; lower leaf sheaths glabrous.

48. Perigynia 2.5–3.5 mm long, nerveless, stipitate; leaves 1.5–3 mm wide .... *C. stylosa*

48. Perigynia 3.5–4.5 mm long, 2-ribbed and strongly nerved, with a broad base; leaves 3–8 mm wide .... *C. raynoldsii*

45. Perigynia flattened.

49. Culms phyllopodic, with many leaves on the lower third and clothed at base with dried-up leaves of previous year, not purplish-tinged at base.

50. Perigynia ovate to orbicular; spikes oblong ...... *C. paysonis*

50. Perigynia narrowly ovate to ovate; spikes cylindric .... *C. microchaeta*

49. Culms aphyllopodic, without leaves on the lower third, leaves of the previous year not persistent or much desiccated at flowering time, purplish-tinged at base.

51. Pistillate scales obtuse to acute, the midvein nearly or quite obsolete or inconspicuous; perigynia two-ribbed, otherwise nerveless; achenes slenderly long-stipitate ... *C. podocarpa*

51. Pistillate scales mucronate or aristate with conspicuous midvein; perigynia marginally two-ribbed and obscurely nerved; achenes short-stipitate.

52. Pistillate scales mucronate or aristate with awns shorter than 1 mm; beaks of perigynia about 0.5 mm long .... *C. spectabilis*

52. Pistillate scales aristate with a slender rough awn 3–12 mm long; beaks of perigynia about 0.25 mm long .. .. *C. macrochaeta*

44. Terminal spike gynaecandrous (pistillate above staminate).

53. Culms lateral, arising singly and not clothed at base with the dried-up leaves of the previous year; the lower sheaths breaking ventrally and conspicuously filamentous .... *C. buxbaumii*

53. Culms central, arising from the midst of the persistent leaves of the previous year and their desiccated fragments; sheaths not filamentous.

54. Pistillate scales abruptly awned .... *C. gmelinii*

54. Pistillate scales obtuse to acute, lacking a distinct awn.

55. Spikes sessile or on short stiff peduncles; peduncles of the lowest spikes shorter than one half of the spike.

56. Lateral spikes cylindrical, separate, with only the upper ones sometimes loosely aggregated into a very much elongated head .... *C. parryana*

56. Lateral spikes ovate or globose, closely aggregated into a subglobose or ovate head.

57. Perigynia flattened; sheaths of the basal leaves with conspicuous black streaks .... *C. albonigra*

57. Perigynia suborbicular; sheaths of the basal leaves brownish .... *C. media*

55. Spikes pedunculate; peduncles of the lowest spikes longer than one half of the spike.

58. Perigynia strongly flattened, papery, 5 mm long, light-green; spikes 6–10, oblong-cylindric, nodding in one direction .... *C. mertensii*

58. Perigynia round or obtusely triangular in cross-section, only slightly flattened, 2.5–4.5 mm long, glaucous dull-green to brownish black; spikes 3–6, shortly-cylindric, erect or spreading in all directions.

59. Perigynia smooth; scales acute to acuminate .... *C. epapillosa*

59. Perigynia papillose or spinulose; scales obtuse.

60. Lateral spikes oblong or ovoid; perigynia sub-inflated, triangular, only scarcely compressed .... *C. atrosquama*

60. Lateral spikes cylindrical to linear; perigynia compressed .... *C. atratiformis*

30. Perigynium beak bidentate.

61. Pistillate spikes close together forming a head below the staminate spike(s), sometimes with one solitary pistillate spike slightly remote or at the base of the culm.

62. Perigynia 7–10 mm long, coarsely nerved; staminate spikes 1–4, 2–6 cm long ... . *C. retrorsa*

62. Perigynia 2–6 mm long, ribbed; staminate spike solitary, 0.5–2 cm long.

63. Fertile culms shorter than 10 cm, usually hidden within stiff gray green leaves; perigynia subcoriaceous, dull gray green .... *C. rugosperma*

63. Fertile culms usually taller than 10 cm, taller than their basal green or yellow green leaves; perigynia membranaceous, green or yellow green.

64. Lower perigynia deflexed; beak curved, about as long as the body .... *C. flava*

64. Lower perigynia not deflexed; beak nearly straight, distinctly shorter than the body .... *C. viridula*

61. Pistillate spikes remote, not forming a head.

65. Sheaths of leaves hairy on the ventral side .... *C. atherodes*

65. Sheaths of leaves glabrous on the ventral side.

66. Beaks of perigynia serrulate on the margin.

67. Terminal spike usually androgynous (staminate above pistillate), the upper lateral spike sessile and staminate, androgynous or pistillate, the lower 1 or 2 spikes pistillate or androgynous; plants with slender horizontal stolons; perigynia ciliate.

68. Perigynia lanceolate, 1.5–1.75 mm wide, tapering to the apex; spikes 2–5, the terminal one with but a few perigynia at base .... *C. petricosa*

68. Perigynia oblong-ovate, 2.0–2.5 mm wide, abruptly minutely beaked; spikes more numerous, the terminal one more strongly pistillate at base .... *C. franklinii*

67. Terminal spike staminate or gynaecandrous (pistillate above staminate), the lateral pistillate; plants densely cespitose or with short ascending stolons; perigynia ciliate-serrulate only on the beak margin.

69. Perigynia narrowly lanceolate, 1 mm wide, flattened-triangular, the upper half very long-tapering .... *C. misandra*

69. Perigynia narrowly lanceolate to ovate, 1.25–2.25 mm wide, contracted into a beak .... *C. ablata*

66. Beaks of perigynia smooth on the margin.

70. Perigynia nerveless.

71. Pistillate spikes sessile, the lowest shortly pedunculate, erect; beak shorter than the perigynium body; the base of culms scaly .... *C. amplifolia*

71. Pistillate spikes pedunculate, the lowest nodding; beak as long as or longer than the perigynium body; the base of culms fibrous .... *C. sprengelii*

70. Perigynia with nerves on both sides.

72. Perigynia finely and closely ribbed with 6–10 ribs on each side; scales with a long rough awn.

73. Teeth of the perigynia beaks straight; perigynia suborbicular in cross-section, more or less inflated, membranaceous or submembranaceous; rootstock with long slender horizontal stolons; ligules short, wider than long or about as wide as long; achenes broadly obovate .... *C. hystricina*

73. Teeth of the perigynia beaks curved out; perigynia flattened-triangular, scarcely inflated, coriaceous, closely or rather closely enveloping achenes below; culms cespitose and not stoloniferous; ligules much longer than wide, very conspicuous; achenes oblong-ovoid or oblong-obovoid .... *C. comosa*

72. Perigynia coarsely ribbed with fewer than 6 nerves on each side; scales acute to attenuate, without a rough awn.

74. Lower perigynia reflexed to widely spreading; bracts several times longer than the inflorescence .... *C. retrorsa*

74. Lower perigynia not reflexed; bracts more than a few times exceeding the inflorescence.

75. Rootstocks short-creeping, lacking long, slender, horizontal stolons; perigynia appressed-ascending, ascending or spreading-ascending, strongly nerved; lower sheaths breaking and filamentose; ligules much longer than wide .... *C. exsiccata*

75. Rootstocks with long, slender, horizontal stolons; perigynia spreading or squarrose at maturity; lower sheaths not breaking and becoming filamentose; ligules wider than long or slightly longer than wide.

76. Leaves rather sparingly septate-nodulose; staminate spikes usually 1, often with 1 or 2 smaller ones at base; perigynia purplish-black, except for the white-hyaline apex; pistillate spikes 0.8–3 cm long; culms 1.5–4.5 dm high .... *C. membranacea*

76. Leaves strongly or very strongly septate-nodulose; staminate spikes 2–4; perigynia yellowish-green, straw-colored, or light brownish; pistillate scales lighter-colored; pistillate spikes 1–15 cm long; culms 3–12 dm high.

77. Plants with channelled leaves 2–4 mm broad; pistillate spikes dense; perigynia rounded ovoid (flask-shaped), horizontally spreading, abruptly contracted into abbreviated beaks; culms obtusely triangular above; rare in BC .... *C. rostrata*[3]

77. Plants with flat leaves 4–7 mm broad; pistillate spikes loose; perigynia obliquely divergent, gradually tapering into abbreviated beaks; culms sharply triangular above; common in BC .... *C. utriculata*

Carex ablata **Bailey** (*C. luzulina* auct. non Olney, *C. luzulina* Olney var. *ablata* [Bailey] F.J. Herm.)
Woodrush sedge
Habitat/Range: Moist meadows and bogs in the subalpine to alpine zones; locally frequent in SW BC; S to ID, UT and CA.

Carex aenea **Fern.**
Bronze sedge
Habitat/Range: Dry, disturbed sites and open forests in the lowland and montane zones; common in BC in and E of the Coast-Cascade Mountains, infrequent in coastal SW BC; N to AK and YT, E to Labr. and NF and S to MA, PA, MN, SD, MT, ID and WA.

Carex albonigra **Mack. in Rydb.**
Two-toned sedge
Habitat/Range: Dry, open slopes in the upper subalpine and alpine zones; common in BC in and E of the Coast-Cascade Mountains, infrequent in coastal SW BC; N to AK, YT and NT, E to SW AB and S to CO, AZ and CA.

Carex amplifolia **Boott in Hook.**
Bigleaf sedge
Habitat/Range: Moist meadows, swamps and bogs in the lowland and montane zones; rare in mainland SW BC; S to ID and CA.

Carex anthoxanthea **Presl**
Yellow-flowered sedge
Habitat/Range: Wet meadows and bogs from the lowland to subalpine zones; frequent in coastal BC; amphiberingian, N to AK, YT and NT; E. Asia.

Carex aperta **Boott in Hook.** (*C. turgidula* Bailey)
Columbia sedge
Habitat/Range: Wet lake margins and meadows in the lowland, steppe and montane zones; frequent in S BC, south of 55°N, absent on the Queen Charlotte Islands; E to SW AB and S to ID, MT and OR.

Carex aquatilis **Wahlenb.**
Water or northern sedge
Habitat/Range: Wet meadows, fens, streambanks and lakeshores in all vegetation zones; common throughout BC, less frequent along the coast; circumpolar, N to AK, YT and NT, E to NF and S to NY, OH, IN, NE, NM and CA; Greenland, Eurasia.

---

3 This species probably does not occur in BC at this time.

Notes: Two subspecies may be recognized in BC:

1. Staminate spikes 2–3; plants generally more than 2 dm tall; plants more common southward .... ssp. *aquatilis*

1. Staminate spikes 1; plants generally less than 2 dm tall; plants more common northward .... ssp. *stans* (Drej.) Hult.

Carex arcta **Boott**
Northern clustered sedge
Habitat/Range: Moist to wet sites in the lowland, steppe and montane zones; common in S BC; N to AK, YT and NT, E to Labr. and NB and S to MI, MN, ID and CA.

Carex atherodes **Spreng.**
Awned sedge
Habitat/Range: Wet meadows, fens, marshes and lakeshores in the montane and steppe zones; frequent in BC south of 53°N and E of the Coast-Cascade Mountains, infrequent northward; circumpolar, N to AK, YT and NT, E to PQ and S to ME, NY, IN, MO, UT and OR; Eurasia.

Carex athrostachya **Olney**
Slender-beaked sedge
Habitat/Range: Moist to wet meadows and open forests in the lowland, montane and steppe zones; common in BC south of 55°N, rare northward and along the coast, absent from the Queen Charlotte Islands; N to AK, E to SK and S to ND, CO and CA.

Carex atratiformis **Britton** (*C. raymondii* Calder, *C. atratiformis* Britt. ssp. *raymondii* [Calder] Porsild)
Black sedge
Habitat/Range: Moist sites in the montane zone; frequent in N BC; N to AK, YT and NT and E to NF.

Carex atrosquama **Mack.** (*C. atrata* auct. non L., *C. atrata* L. spp. *atrosquama* [Mack.] Hult.)
Black-scaled sedge
Habitat/Range: Moist to mesic slopes in the subalpine and alpine zones; common in BC in and east of the Coast-Cascade Mountains; N to AK, YT and NT, E to SW AB and S to CO, ID and OR.

Carex aurea **Nutt.**
Golden sedge
Habitat/Range: Moist to wet meadows and bogs in all vegetation zones; frequent in BC east of the Coast-Cascade Mountains; N to AK, YT and NT, E to NF and S to PA, NE, NM and CA.

Carex bebbii **(Bailey) Fern.**
Bebb's sedge
Habitat/Range: Wet meadows and streamsides in the lowland, steppe vegetation and montane zones; common in mainland BC south of 57°N, rare northward, absent from Vancouver Island and the Queen Charlotte Islands; N to AK and NT, E to NF and S to NJ, NB, CO and OR.

Carex bicolor **Bell. ex All.**
Two-coloured sedge
Habitat/Range: Moist to wet meadows and shorelines in the montane to alpine zones; rare in N BC; circumpolar, N to AK, YT and NT and E to NF, Greenland, Iceland, Eurasia.

Carex bigelowii **Torr. in Schwein.**
Bigelow's or stiff sedge
Habitat/Range: Wet margins of streams and pools in the subalpine and alpine zones; infrequent in N BC; circumpolar, N to AK, YT and NT, E to NF and S to ME and NY, Greenland, Iceland, Eurasia.

***Carex brevicaulis* Mack.** (*C. deflexa* Hornem. var. *brevicaulis* [Mack.] Boivin)
Short-stemmed sedge
Habitat/Range: Dry grassy slopes in the lowland zone; locally frequent in SW BC; S to CA.
Notes: This species is part of the poorly studied *C. deflexa/C. rossii* complex.

***Carex brevior* (Dewey) Mack.**
Short-beaked sedge
Habitat/Range: Moist to dry sites in the steppe and montane zones; infrequent in SC and SE BC; E to PQ and S to TN, AR, TX, AZ and OR.

***Carex brunnescens* (Pers.) Poir** (*C. brunnescens* ssp. *alaskana* Kalela and ssp. *pacifica* Kalela, *C. sphaerostachya* [Tuckerm.] Dewey, *C. brunnescens* var. *sphaerostachya* [Tuckerm.] Kukenth.)
Brownish sedge
Habitat/Range: Bogs, fens and wet sites in forests in the montane and subalpine zones; frequent in N BC, rare in mountains S of 54°N.; circumpolar, N to AK, YT and NT, E to NF and S to NJ, NC, GA, CO, UT and OR; Eurasia.

***Carex buxbaumii* Wahl.** (*C. adelostoma* Krecz., *C. canescens* auct. non L.)
Buxbaum's or club sedge
Habitat/Range: Wet bogs, fens, wet shrub lands, shorelines and meadows in the lowland and montane zones; frequent in WC and S BC, absent from N BC except the extreme NW; circumpolar, N to AK and YT, E to NF and S to GA, AR, CO, UT and CA; Eurasia, Australia.

***Carex canescens* L.** (*C. curta* Good, *C. cinerea* Poll.; *C. arctaeformis* Mack. = ssp. *arctaeformis*)
Grey or white sedge
Habitat/Range: Bogs, swamps, fens, wet meadows and streambanks in all but the alpine zone; common in S BC, less frequent northward; circumpolar, N to AK, YT and NT, E to Labr. and NF and S to VA, OH, IN, AZ and CA; Greenland, Iceland, Eurasia, S America, Australia, New Guinea.
Notes: Two subspecies can be distinguished in BC:

1. Scales smaller than perigynia, perigynia exposed, conspicuous in the spikes; lower spikes remote .... ssp. *canescens*

1. Scales enveloping perigynia towards the base, perigynia concealed; spikes all closely approximate .... ssp. *arctaeformis* (Mack.) Calder & Taylor

***Carex capillaris* L.** (*C. chlorostachys* Stevens, *C. fuscidula* Krecz. ex Egor.)
Hairlike or hair sedge
Habitat/Range: Moist to wet shorelines, bogs, fens and meadows in the montane to alpine zones; frequent throughout BC east of the Coast-Cascade Mountains; circumpolar, N to AK, YT and NT, E to Labr. and NF and S to VT, MI, MN, UT, NM, NV and OR; Greenland, Iceland, Eurasia.

***Carex capitata* L.** (*C. arctogena* H. Smith)
Capitate sedge
Habitat/Range: Wet to dry sites in the montane to alpine zones; frequent throughout BC in and E of the Coast-Cascade Mountains; circumpolar, N to AK, YT and NT, E to NF and S to NH, UT, NM, CA and MX; Greenland, Iceland, Eurasia, disjunct to S South America.

***Carex chordorrhiza* Ehrh. ex L. f.**
Cordroot sedge or string sedge
Habitat/Range: Wet bogs and fens in the montane zone; common in N and C BC, rare southward; circumpolar, N to AK, YT and NT, E to NF and S to VT, IN and IA; Iceland, Eurasia.

Carex circinata **C.A. Meyer**
Coiled sedge
Habitat/Range: Moist cliffs in the subalpine and alpine zones; infrequent in south-coastal BC, frequent in the Queen Charlotte Islands and adjacent mainland, rare in the Coast-Cascade Mountains; N to AK and S to NW WA.

Carex comosa **Boott**
Bearded sedge
Habitat/Range: Shorelines and wet meadows in the lowland, steppe and montane zones; rare in SW and SC BC; E to PQ, NB and NS and S to AR, SC, GA, FL, ID and CA.

Carex concinna **R. Br.**
Low northern sedge
Habitat/Range: Moist to dry forests in the montane zone; common throughout BC E of the Coast-Cascade Mountains; N to AK, YT and NT, E to NF and S to MI, SD, CO and OR.

Carex concinnoides **Mack.**
Northwestern sedge
Habitat/Range: Moist to dry forests in all but the montane to alpine zone; common in SC and SE BC east of the Coast-Cascade Mountains, less frequent northward to 56°N; E to SW AB and S to MT, ID and N CA.

Carex crawei **Dewey**
Crawe's sedge
Habitat/Range: Moist meadows in the montane zone; rare in SE BC; E to NF and S to NJ, AL, KS, UT and WA.

Carex crawfordii **Fern.**
Crawford's sedge
Habitat/Range: Moist to wet sites in the lowland, steppe and montane zones; common throughout BC north to 56°N, rare northward and on the W slopes of the Coast-Cascade Mountains; N to AK, E to NF, and S to NJ, MI, TN, MT, ID and WA.

Carex cusickii **Mack. in Piper & Beattie** (*C. diandra* Schrank var. *ampla* [Bailey] Kukenth.)
Cusick's sedge
Habitat/Range: Wet bogs, fens, marshes and lake margins in the lowland and montane zones; frequent in S BC, less frequent N to 55°N; S to UT and CA.

Carex deflexa **Hornem.**
Bent sedge
Habitat/Range: Forest clearings, open spaces in the montane and subalpine zones; frequent in BC east of the Coast-Cascade Mountains, infrequent on the coast; N to AK and YT, E to NF and S to MA and MI.
Notes: Closely related to *C. rossii*, these two taxa are sometimes considered conspecific. In that case *C. deflexa* (the oldest name in the complex) is used in a broad sense (e.g., Taylor 1983).

Carex deweyana **Schwein.** (*C. bolanderi* Olney = var. *bolanderi*; *C. leptopoda* Mack. in Rydb. = var. *leptopoda*)
Dewey's sedge
Habitat/Range: Wet places in alluvial forests, forest margins and clearings in the lowland, steppe, and montane zones; common in BC south of 55°N, infrequent northward; E to NF and S to PA, MI, SD, CO, NM, AZ and CA.
Notes: Variable species with three varieties in BC. Sometimes these are treated as separate species.

1. Rootstocks rarely elongate, the culms usually densely cespitose; perigynia rounded at the base, 4.5–5.5 mm long, beak shallowly bidentate, the upper part of the body covered by the translucent scale; achenes 2.0–2.5 mm long .... var. *deweyana*

1. Rootstocks slender, elongate; perigynia tapering at the base, stipitate, 3.5–4.5 mm long; achenes 1.5–1.8 mm long.

   2. Beaks of perigynia shallowly bidentate, the upper part of the bodies not covered by the scales; spikes ovoid or oblong; scales not reddish brown-tinged .... var. *leptopoda* (Mack. in Rydb.) Boivin

   2. Beaks of perigynia deeply bidentate, the bodies covered by the scales; spikes linear-oblong; scales usually reddish brown-tinged .... var. *bolanderi* (Olney) Boott

***Carex diandra* Schrank**

Lesser panicled or lesser tussock sedge

Habitat/Range: Bogs, swamps and fens in all but the alpine zone; common in BC in and east of the Coast-Cascade Mountains, rare in SW BC; circumboreal, N to AK, YT and NT, E to NF and S to NJ, IN, NE, MO, CO, UT and CA; Iceland, Eurasia.

***Carex disperma* Dewey** (*C. tenella* Schukuhr)

Soft-leaved sedge

Habitat/Range: Bogs, swamps, wet meadows and moist forests in all but the alpine zone; common in BC east of the Coast-Cascade Mountains, rare in coastal BC; circumpolar, N to AK, YT and NT, E to NF and S to NJ, PA, MN, SD, UT, AZ NM and CA; Eurasia.

***Carex douglasii* Boott**

Douglas' sedge

Habitat/Range: Dry sites in the steppe and montane zones; infrequent in SC and SE BC; E to S MB and S to IA, NE, NM and CA.

***Carex eburnea* Boott in Hook.**

Bristle-leaved sedge

Habitat/Range: Dry, often calcareous, gravelly or rocky sites in the montane zone; frequent throughout E BC (Rocky Mountains); N to AK, YT and NT, E to NF and S to VA, AL, TX, MO and NE.

***Carex echinata* Murr.** (*C. angustior* Mack. in Rydb., *C. cephalantha* Bickn., *C. muricata* auct. non L., *C. stellulata* Good. = ssp. *echinata*; *C. phyllomanica* = ssp. *phyllomanica*)

Star sedge (ssp. *echinata*), coastal stellate sedge (ssp. *phyllomanica*)

Habitat/Range: Wet bogs, sedge meadows or sandy lakeshores and streamsides in the lowland and montane zones; ssp. *echinata* - frequent in coastal BC, rare E of the Coast-Cascade Mountains N to 54°N, ssp. *phyllomanica* - common in coastal BC; ssp. *echinata* - circumpolar, N to AK (Aleutian Islands), E to NF and S to SC, IN, IA, MN, CO, WY, UT, ID and CA; Eurasia; ssp. *phyllomanica* - N to AK and S to CA.

Notes: Two subspecies are recognized for BC.

1. Pistillate scales obtuse, chestnut brown-tinged with broad bright white margins and apex, not sharply keeled, the midvein obscure at apex ...... ssp. *phyllomanica* W. Boott in S. Wats.

1. Pistillate scales acute to cuspidate, yellowish brown-tinged, the margins and apex narrow, opaque or dull whitish, the midvein sharply defined, prominent to apex .... ssp. *echinata*

***Carex eleocharis* Bailey** (*C. duriuscula* C.A. Meyer, *C. stenophylla* Wahl. var. *enervis* [C.A. Meyer] Kukendh.)
Narrow-leaved sedge
Habitat/Range: Moist to dry open sites in the steppe and montane zones; infrequent in C and SE BC, rare in NE BC; N to AK, YT and NT, E to MB and S to IA, KS, NM, AZ, ID and OR.
Notes: Not easily distinguished from the Eurasian *C. stenophylla* and possibly better treated as a subspecies of the latter.

***Carex enanderi* Hult.** (*C. eurystachya* F.J. Herm., *C. lenticularis* var. *dolia* [M.E. Jones] Standley)
Enander's sedge
Habitat/Range: Wet margins of streams and pools in the subalpine zone; infrequent in NW BC; N to AK.

***Carex engelmannii* Bailey** (*C. breweri* Boott var. *paddoensis* [Suks.] Cronq.)
Engelmann's sedge
Habitat/Range: Moist to dry meadows and talus slopes in the upper subalpine and alpine zones; locally frequent in SC BC; S to WA, disjunct to MT, WY, CO, UT and CA.
Notes: Further study may indicate that this species is better treated as a variety of *C. breweri*.

***Carex epapillosa* Mack. in Rydb.** (*C. atrata* L. var. *erecta* Boott in S. Wats., *C. heteroneura* var. *epapillosa* [Mack.] F.J. Herm)
Blackened sedge
Habitat/Range: Moist to dry meadows in the upper subalpine to alpine zones; rare in SC BC; E to SW AB and S to CO, NV and CA.

***Carex exsiccata* Bailey** (*C. vesicaria* auct. non L., *C. vesicaria* L. var. *major* Boott)
Inflated sedge
Habitat/Range: Bogs, swamps, marshes, shorelines and wet meadows in the lowland, steppe and montane zones; common in S BC, less frequent northward; N to S AK, E to SK and S to MT and CA.
Notes: Closely related to the European and eastern North American *C. vesicaria*.

***Carex feta* Bailey** (*C. straminea* Willd. var. *mixta* Bailey)
Green-sheathed sedge
Habitat/Range: Ditches, marshes and wet meadows in the lowland and montane zones; rare in extreme S BC, S to CA.

***Carex filifolia* Nutt.**
Thread-leaved sedge
Habitat/Range: Dry open sites in the steppe and montane zones; frequent in SC BC, less frequent eastward and northward; N to AK, YT and NT, E to SW MB and S to MN, NE, TX, NM, AZ and CA.

***Carex flava* L.**
Yellow or large yellow sedge
Habitat/Range: Bogs, swamps, shorelines and wet, sandy sites in the lowland and montane zones; frequent in S BC, rare northward, absent on the Queen Charlotte Islands; disjunct circumpolar, E to NF and S to NJ, PA, IN, MT and ID, Iceland, Europe.

***Carex foenea* Willd.** (*C. siccata* Dewey)
Hay or silvertop sedge
Habitat/Range: Dry, often sandy sites in the steppe and montane zones; frequent in BC east of the Coast-Cascade Mountains except rare in NW BC; N to S YT and to NT, E to S PQ and S to NJ, OH, NE, UT, NM and AZ.

Carex franklinii **Boott in Hook.** (*C. petricosa* Dewey var. *franklinii* [Boott in Hook.] Boivin)
Franklin's sedge
Habitat/Range: Calcareous meadows and rocky slopes in the alpine zone; rare in N BC; known only from Mi. 390 AK Highway; N to AK and YT and E to SW AB.
Notes: Closely related to *C. petricosa*.

Carex garberi **Fern.** (*C. hassei* auct. non Bailey = ssp. *bifaria*)
Garber's sedge
Habitat/Range: Wet meadows and seepage sites in the lowland, steppe and montane zones; infrequent on Vancouver Island (ssp. *bifaria*), frequent E of the Coast-Cascade Mountains (ssp. *garberi*); N to YT, E to PQ, NB and S to ME, PA, MI, WI, UT, NV, and CA.
Notes: Two subspecies (sometimes treated as separate species) can be distinguished:

1. Terminal spike usually gynaecandrous (pistillate above staminate) with at least 1/4 of the flowers pistillate and staminate portion 1.2–2.7 mm wide; lateral spikes on average with less than 20 flowers .... ssp. *garberi*
1. Terminal spike staminate, rarely with a few pistillate flowers at the apex, 1.8–3 mm wide; lateral spikes on average with 20 or more flowers .... ssp. *bifaria* (Fern.) Hult.

Carex geyeri **Boott**
Elk or Geyer's sedge
Habitat/Range: Dry meadows and open forests in the montane to alpine zones; rare in SE BC; E to SW AB and S to CO, UT and CA.

Carex glacialis **Mack.**
Glacier sedge
Habitat/Range: Dry open sites from the montane to alpine zones; infrequent in extreme N BC; circumpolar N to AK, YT and NT and E to NF; Greenland, Iceland, N. Eurasia.

Carex glareosa **Wahlenb. var. *amphigens*** **Fern.** (*C. marina* auct. non Dewey)
Lesser saltmarsh sedge
Habitat/Range: Saline marshes in the lowland zone; rare in coastal areas N of Vancouver Island; circumpolar, N to AK, disjunct to coastal PQ and NF; Eurasia.

Carex gmelinii **Hook & Arn.**
Gmelin's sedge
Habitat/Range: Saline marshes, bluffs and meadows in the lowland zone; rare in coastal areas S to the N tip of Vancouver Island; amphiberingian, N to AK; E Asia.

Carex gynocrates **Wormsk. ex Drej.** (*C. dioica* L. ssp. *gynocrates* [Wormsk. ex Drej.] Hult.)
Yellow bog sedge
Habitat/Range: Bogs, swamps, fens and wet meadows in the upper montane to alpine zones; common in and E of the Coast-Cascade Mountains throughout BC; amphiberingian, N to AK, YT and NT, E to NF and S to MA, PA, MN, MI, CO and OR; Greenland, E. Asia.

Carex haydeniana **Olney in S. Wats.** (*C. macloviana* d'Urv. ssp. *haydeniana* [Olney in S. Wats.] Tayl. & MacBryde)
Hayden's sedge
Habitat/Range: Moist to mesic meadows, rocky slopes and moraines in the subalpine and alpine zones; infrequent in SC and SE BC; E to SW AB and S to WA, ID and MT.

Carex heleonastes **Ehrh. in L. f.**
Hudson Bay sedge
Habitat/Range: Bogs and fens in the montane zone; rare in C and SC BC; circumpolar, N to AK, YT and NT and E to ON and NF; Eurasia.

***Carex hendersonii* Bailey**

Henderson's sedge

Habitat/Range: Moist forests in the lowland zone; locally frequent on SE Vancouver Island, the Gulf Islands and the adjacent mainland; S to CA.

***Carex hindsii* Clarke** (*C. lenticularis* Michx. var. *limnophila* [Holm.] Cronq. in Hitchc. *et al.*)

Hinds' sedge

Habitat/Range: Marshes and wet meadows in the lowland zone; frequent along the coast; N to AK and S to NW CA.

Notes: Some authors (e.g., Calder and Taylor 1968, Taylor 1983) treat this species as synonomous with *C. kelloggii* while others (e.g., Hitchcock *et al.* 1969) treat both as varieties of *C. lenticularis.*

***Carex hoodii* Boott in Hook.**

Hood's sedge

Habitat/Range: Moist to dry meadows and forest openings in all vegetation zones; common in BC south of 56°N except absent from the Queen Charlotte Islands, N Vancouver Island and the adjacent mainland; E to SW SK and S to SD, CO, UT and CA.

***Carex hystricina* Muhl. ex Willd.**

Porcupine sedge

Habitat/Range: Swamps, shorelines and wet meadows in the steppe and montane zones; rare, scattered in BC east of the Coast-Cascade Mountains and S of 56°N; E to NF and S to VA, TN, TX, NM, AZ and CA.

***Carex illota* Bailey**

Sheep sedge

Habitat/Range: Bogs, shorelines and wet meadows in the upper montane to alpine zones; frequent in S BC; E to SW AB and S to CO, UT and CA.

***Carex inops* Bailey** (*C. pensylvanica* auct. non Lam., *C. pensylvanica* Lam. var. *vespertina* Bailey = ssp. *inops*, *C. pensylvanica* Lam. var. *digyna* Boeckl. = ssp. *heliophila*)

Long-stoloned sedge

Habitat/Range: Dry sites in the lowland, montane and steppe zones; infrequent in SW and SC BC (ssp. *inops*), and in NE BC (ssp. *heliophila*); S to NM and N CA.

Notes: Two subspecies occur in BC:

1. Lower pistillate spike elongate, with a short peduncle, its bract often short-sheathing; peduncles of staminate spike 4–10 mm long .... ssp. *inops*

1. Lower pistillate spike short, sessile, its bract sheathless; peduncles of staminate spike 0.8–5 mm long .... ssp. *heliophila* (Mack.) Crins

***Carex interior* Bailey**

Inland sedge

Habitat/Range: Bogs, fens, swamps and wet sites in the lowland, steppe and montane zones; frequent E of the Coast-Cascade Mountains, infrequent in W BC; N to AK, YT and NT, E to NF and S to PA, IN, KS, MX and CA.

***Carex interrupta* Boeckl.**

Green-fruited sedge

Habitat/Range: Streamsides and wet places in the lowland zone; rare in SW BC, known only from Duncan and Chilliwack; S to S OR.

***Carex kelloggii* Boott in S. Wats.** (*C. lenticularis* auct. non Michx., *C. lenticularis* Michx. var. *lipocarpa* [Holm] Standley)
Kellogg's sedge
Habitat/Range: Swamps, bogs and wet sites in all but the alpine zone; common throughout BC S of 56°N, rare in NW BC; E to SW AB and S to CO, UT and CA.

***Carex krausei* Boeckl.** (*C. capillaris* L. ssp. *krausei* [Boeckl.] Böcher and var. *krausei* [Boeckl.] Crantz)
Krause's sedge
Habitat/Range: Wet gravel banks in the subalpine zone; rare in NW BC; disjunct circumpolar, N to AK, YT, and NT, and E to MB, PQ and NF; Greenland, Iceland, N Asia.

***Carex lachenalii* Schkuhr** (*C. bipartita* Bell. ex All., *C. tripartita* auct. non All., *C. lagopina* Wahlenb.)
Two-parted or hare's-foot sedge
Habitat/Range: Mesic open rocky slopes in the subalpine and alpine zones; common throughout BC in and E of the Coast-Cascade Mountains; circumpolar, N to AK, YT and NT, E to NF and S to CO and UT; Greenland, Eurasia, New Zealand.
Notes: Nomenclature of this species is confused. In our selection of the name we decided to follow the majority of European literature.

***Carex laeviculmis* Meinsh.**
Smooth-stemmed sedge
Habitat/Range: Bogs, shorelines, streambanks, wet meadows and moist forests in the lowland and montane zones; common in BC south of 55°N, rare in N BC; amphiberingian N to AK and S to ID, MT and CA; E Asia.

***Carex lanuginosa* Michx.** (*C. lasiocarpa* Ehrh. var. *latifolia* [Boeckl.] Gleason)
Woolly sedge
Habitat/Range: Ditches, streambanks, swamps, marshes and wet meadows in the lowland, steppe and montane zones; common in S BC, absent on N Vancouver Island, the Queen Charlotte Islands and adjacent mainland; E to NF and S to TN, TX, NM and CA.

***Carex lapponica* O. Lang** (*C. canescens* L. var. *subloliacea* Laest.)
Lappland sedge
Habitat/Range: Wet peat bogs and open peat bog forests in the montane zone; rare in NW BC; circumpolar, N to AK and YT; Eurasia.

***Carex lasiocarpa* Ehrh.** (*C. lasiocarpa* ssp. *americana* (Fern.) Löve & Bernard)
Slender sedge
Habitat/Range: Bogs, swamps, shorelines and wet meadows in the lowland, steppe and montane zones; frequent in S and SC BC, rare northward, absent on N Vancouver Island, the Queen Charlotte Islands and adjacent mainland; circumpolar, N to AK, YT and NT, E to NF and S to NJ, IA, OH, ID and WA; Eurasia.

***Carex lenticularis* Michx.** (*C. enanderi* Hult., *C. eurystachya* F.J. Herm. and *C. plectocarpa* F.J. Herm. = var. *dolia*, *C. hindsii* Clarke = var. *limnophila*, *C. kelloggii* W. Boott in S. Wats. = var. *lipocarpa*)
Lens-fruited sedge
Habitat/Range: Marshes and wet meadows in the lowland zone (var. *limnophila*), streamsides and ponds in the subalpine zone (var. *dolia*) and bogs and wet sites in all but the alpine zone (var. *lipocarpa*); var. *dolia* - infrequent in W and SE BC, var. *limnophila* - frequent along the coast, and var. *lipocarpa* - common throughout BC south of 56°C, rare in NW BC; var. *dolia* - N to SE AK, E to SW AB and S to NW MT, var. *limnophila* - N to SE AK and S to NW CA, and var. *lipocarpa* - N to SE AK and S to CO, WY, NM, AZ and CA.
Notes: Three varieties occur in BC.[4]

---

4 Key from Standley (1985).

1. Terminal spike pistillate with only a few staminate flowers basally; basal spike up to 1.5 cm long, with a peduncle up to 1 cm long .... var. *dolia* (M.E. Jones) Standley
1. Terminal spike staminate; basal spike up to 5 cm long, with a peduncle up to 5 cm long.
    2. Pistillate scales light brown; beaks of perigynia green; auricles of involucral bracts white .... var. *lenticularis*[5]
    2. Pistillate scales purplish brown or blackish; beaks of perigynia purplish brown; auricles of involucral bracts generally purplish brown.
        3. Basal spikes 4–6 mm wide; stipes of perigynia 0.4–0.7 mm long; inflorescences compact .... var. *limnophila* (Holm) Cronq. in C.L. Hitchc. *et al.*
        3. Basal spikes 3–4 mm wide; stipes of perigynia up to 0.4 mm long; inflorescences diffuse.
            4. Perigynia with 3 nerves on the abaxial face, nerves sometimes present but not well-developed adaxially; beaks of the perigynia generally purplish brown for entire length; perigynia green, often with purplish brown spots distally .... var. *impressa* (Bailey) Standley[5]
            4. Perigynia with 5–7 strong nerves on both faces; beaks of the perigynia generally purplish brown only at the apex; perigynia uniformly green .... var. *lipocarpa* (Holm) Standley

Carex leptalea **Wahlenb.** (*C. jamescalderi* Boivin = ssp. *pacifica*)
Bristle-stalked sedge
Habitat/Range: Bogs, swamps, shorelines and fens in the lowland, steppe and montane zones; frequent throughout BC west (ssp. *pacifica*) and east (ssp. *leptalea*) of the Coast-Cascade Mountains; N to AK, YT and NT, E to NF and S to TX, FL, CO and CA.
Notes: Two subspecies can be distinguished in BC:

1. Pistillate scales obtuse; perigynia smaller, 2.5–3.5 mm long; E of the Coast-Cascade Mountains .... ssp. *leptalea*
1. Pistillate scales acuminate or aristate; perigynia 3.5–5 mm long; W of the Coast-Cascade Mountains .... ssp. *pacifica* Calder & Taylor

Carex limosa **L.**
Shore or mud sedge
Habitat/Range: Bogs, shorelines and fens in the lowland, montane and subalpine zones; frequent throughout BC except absent on the Queen Charlotte Islands; circumpolar, N to AK, YT and NT, E to NF and S to NJ, MN, UT and CA; Iceland, Eurasia.

Carex livida **(Wahl.) Willd.** (*C. livida* var. *grayana* Fern.).
Livid sedge
Habitat/Range: Bogs, swamps, wet meadows and moist forests in the lowland and montane zones; common along the coast, less frequent in E BC (Rocky Mountains); disjunct circumpolar, N to AK and S YT, E to NF and S to NJ, MN, ID, disjunct in NW CA; Iceland, N Europe and E Asia.

Carex loliacea **L.**
Ryegrass sedge
Habitat/Range: Marshes, streambanks, fens and forests in the montane zone; infrequent in BC north of 51°N; circumpolar, N to AK, YT, and NT and E to N MN; Eurasia.

---

5 These varieties have yet to be collected in BC. They are likely to be found in the future since they occur only a few km south of the border either in NW WA (var. *impressa*) or NW MT (var. *lenticularis*).

***Carex lyngbyei* Hornem.** (*C. cryptocarpa* C.A. Meyer., *C. lyngbyei* var. *robusta* [Bailey] Cronq.)
Lyngbye's sedge
Habitat/Range: Saltmarshes, estuaries and wet saline meadows in the lowland zone; common along the coast; amphiberingian, N to AK and S to CA; coastal E Asia.

***Carex macloviana* D'Urv.**
Falkland Island sedge
Habitat/Range: Lakeshores, ditches and moist to mesic meadows and forests openings in the montane to alpine zones; frequent in N BC, rare in S BC in and east of the Coast-Cascade Mountains; disjunct circumpolar, N to AK, YT and NT, disjunct to Labr., S to WY and CO; Greenland, Iceland and N Europe, S Chile, Falkland Islands.

***Carex macrocephala* Willd. in Spreng.** (*C. anthericoides* Presl, *C. macrocephala* ssp. *anthericoides* [Presl] Hult.)
Large-headed sedge
Habitat/Range: Sand beaches and dunes in the lowland zone; infrequent in coastal BC; amphiberingian, N to AK and S to OR; Japan, China.

***Carex macrochaeta* C.A. Meyer**
Large-awned sedge
Habitat/Range: Moist to wet beaches and meadows in the lowland and montane zones; common in coastal BC; amphiberingian, N to AK and S to WA; NE Asia, Japan.

***Carex magellanica* Lam. ssp. *irrigua* (Wahlenb.) Hult.** (*C. irrigua* Wahlenb., *C. paupercula* Michx.)
Poor, tall bog or bog sedge
Habitat/Range: Bogs, fens and swamps in the lowland, montane and subalpine zones; infrequent throughout BC except absent on the Queen Charlotte Islands; circumpolar, N to AK, YT and NT, E to NF and S to MA, MN, CO, UT and WA; Greenland, Iceland, Eurasia.

***Carex maritima* Gunn.** (*C. incurviformis* Mack. in Rydb., *C. maritima* Gunn. var. *incurviformis* [Mack.] Boivin)
Curved-spiked sedge
Habitat/Range: Mesic meadows and rocky slopes in the alpine zone; rare in SC and SE BC; E to AB and S to CO.

***Carex media* R. Br. in Richards.** (*C. angarea* Steud., *C. norvegica* auct. non Retz., *C. norvegica* Retz. ssp. *inferalpina* [Wahlenb.] Hult., *C. norvegica* Retz ssp. *inserrulata* Kalela)
Scandinavian or alpine sedge
Habitat/Range: Moist to wet meadows and wet shrub lands in all but the lowland zone; common throughout BC in and E of the Coast-Cascade Mountains; circumpolar, N to AK, YT and NT, E to MB and S to UT, NM and WA; Eurasia.
Notes: Western taxonomists have usually included this taxon with *C. norvegica* Retz, a species of eastern N. America and the southern Rocky Mountains (Murray 1969).

***Carex membranacea* Hook. in Parry**
Fragile sedge
Habitat/Range: Bogs, shorelines and wet meadows in the montane and subalpine zones; rare in extreme N BC; amphiberingian, N to AK, YT and NT and E to PQ; E Asia.

***Carex mertensii* Prescott in Bong.**
Merten's sedge
Habitat/Range: Roadsides, ditches, and wet meadows in the lowland and montane zones; common in BC south of 55°N, rare in NW BC; amphiberingian, N to AK and S YT, E to AB and S to ID, MT and CA; E Asia.

Carex microchaeta **Holm**

Small-awned sedge

Habitat/Range: Moist to wet meadows in the montane to alpine zones; common in extreme N BC; N to AK, YT and NT.

Carex microglochin **Wahlenb.**

Few-seeded bog or bristle sedge

Habitat/Range: Shorelines, streambanks, calcareous fens and wet meadows in the montane and subalpine zones; infrequent in BC east of the Coast-Cascade Mountains and N of 52°N; circumboreal, N to AK, YT and NT and E to NF; Greenland, Iceland, Eurasia, S South America.

Carex microptera **Mack.**

Small-winged sedge

Habitat/Range: Bogs, swamps, lakeshores, ditches, wet meadows and moist forests in the montane to alpine zones; frequent throughout BC east of the Coast-Cascade Mountains; N to S YT and NT, E to MB and S to SD, AZ, NM and CA.

Carex misandra **R. Br.** (*C. fuliginosa* Schkuhr ssp. *misandra* [R. Br.] Nyman)

Short-leaved sedge

Habitat/Range: Dry slopes in the alpine zone; rare in NE BC; circumpolar, N to AK, YT and NT, E to Labr. and disjunct to WY, CO and UT; Greenland, N Eurasia.

Carex nardina **Fries** (*C. hepburnii* Boott)

Spikenard sedge

Habitat/Range: Dry slopes in the alpine zone; frequent in and E of the Coast-Cascade Mountains, rare on Vancouver Island; disjunct circumpolar, N to AK, YT and NT, E to NF and S to CO, MN and OR; Greenland, Iceland, N Europe, E. Asia.

Carex nigricans **C.A. Meyer**

Black alpine sedge

Habitat/Range: Moist to wet meadows, margins of streams and snowbed sites in the montane to alpine zones; common in BC south of 54°N, less frequent northward; N to YT, E to SW AB and S to CO, UT and CA.

Carex obnupta **Bailey** (*C. magnifica* Dewey)

Slough sedge

Habitat/Range: Swamps, marshes, riverbanks and wet meadows and forests in the lowland zone; common in coastal BC; S to CA.

Carex obtusata **Lilj.**

Blunt sedge

Habitat/Range: Dry, open sites from the steppe to alpine zones; common in SC BC (Cariboo region), frequent in N BC; circumpolar, N to AK, YT and NT, E to W ON, and S to MN, NE, SD, CO, UT and NM; Eurasia.

Carex pachystachya **Cham. ex Steud.**

Thick-headed sedge

Habitat/Range: Mesic to wet sites in all vegetation zones; common throughout all but NE BC; amphiberingian, N to AK, YT and NT, E to SK and S to WY, CO and NV; NE Asia.

Carex pallescens **L.**

Pale sedge

Habitat/Range: Grassy meadows and ditches in the lowland zone; rare in SW BC, known only from Hornby Island; introduced from Europe or E North America.

Carex pansa **Bailey** (*C. arenicola* auct. non Schmidt, *C. arenicola* Schmidt ssp. *pansa* [Bailey] Koyama & Calder)
Sand-dune sedge
Habitat/Range: Sandy beaches, dunes and rocky shores in the lowland zone; rare on the Queen Charlotte Islands and W Vancouver Island; S to CA.

Carex parryana **Dewey**
Parry's sedge
Habitat/Range: Moist, often alkaline, meadows and shorelines in the steppe and montane zones; frequent in SC and SE BC; N to AK, YT and NT, E to MB and S to CO, UT and ID.

Carex pauciflora **Lightf.**
Few-flowered sedge
Habitat/Range: Bogs and fens in the lowland, montane and subalpine zones; common along the coast and in C BC, less frequent in N and SC and SE BC; circumpolar, N to AK and YT, E to NF and S to MN, IN, PA and NW WA; Eurasia.

Carex paysonis **Clokey**
Payson's sedge
Habitat/Range: Moist meadows and rocky slopes in the subalpine and alpine zones; rare in extreme SE BC; E to SW AB and S to WY, UT, ID and E OR.

Carex peckii **Howe in Peck** (*C. nigromarginata* Schwein. var. *elliptica* [Boott] Gleason)
Peck's sedge
Habitat/Range: Dry to wet forests in the montane zone; locally frequent in C and E BC north of 52°N; E to PQ, and S to TN, FL, OK and SD.

Carex pedunculata **Willd.**
Peduncled sedge
Habitat/Range: Mesic sites in the montane zone; rare in SE BC, known only from Grandby River; disjunct E to MB, from ON to NF and S to IA, GA and SD.

Carex petasata **Dewey** (*C. liddonii* Boott in Hook.)
Pasture or broad-winged sedge
Habitat/Range: Dry to moist meadows and open forests in the steppe and montane zones; common in SC BC, less frequent elsewhere E of the Coast-Cascade Mountains, ; N to SW YT, E to SW SK and S to CO, UT, NV and OR.

Carex petricosa **Dewey**
Rock-dwelling sedge
Habitat/Range: Dry, gravelly slopes and rock outcrops in the alpine zone; rare in NE BC, known only from Summit Pass; N to AK, YT and NT and E to SW AB.

Carex phaeocephala **Piper**
Dunhead sedge
Habitat/Range: Mesic to dry meadows and rocky slopes in the subalpine and alpine zones; common in SC and SE BC, less frequent northward and on the SW coast, rare on S Vancouver Island, absent on the Queen Charlotte Islands and adjacent mainland; N to AK, YT and SW NT, E to SW AB and S to CO, UT, NV and CA.

Carex pluriflora **Hult.** (*C. rariflora* [Wahlenb.] Smith ssp. *pluriflora* [Hult.] Egor., *C. stygia* auct. non Fr.)
Many-flowered sedge
Habitat/Range: Bogs, marshes, streambanks and wet meadows in the lowland and montane zones; common in coastal BC; amphiberingian, N to AK and S to NW WA; coastal E Asia.

Carex podocarpa **R. Br. in Richards.** (*C. beringensis* C.B. Clarke, *C. montanensis* Bailey)
Graceful mountain sedge
Habitat/Range: Mesic to wet meadows in the subalpine and alpine zones; infrequent in N BC; amphiberingian, N to AK, YT and NT, E to W AB and S to MT, ID and E OR; E Asia.

Carex praegracilis **Boott**
Field sedge
Habitat/Range: Moist to wet, often alkaline, meadows in the montane zone; common in SC and SE BC east of the Coast-Cascade Mountains, rare northward; N to SW YT, E to S MB and S to IA, OK, MN, TX, MX and CA.

Carex prairea **Dewey**
Prairie sedge
Habitat/Range: Bogs, fens and wet meadows in the montane zone; locally frequent in SC and C BC; E to PQ and NS and S to OH, NJ and NB.

Carex praticola **Rydb.**
Meadow sedge
Habitat/Range: Swamps, streambanks and moist to wet meadows and forests in the lowland, steppe and montane zones; common throughout BC except absent on the Queen Charlotte Islands; N to AK, YT and NT, E to NF and S to MA, MI, ND, CO and CA.

Carex preslii **Steud.**
Presl's sedge
Habitat/Range: Mesic to moist grassy slopes in the subalpine zone; infrequent in S BC; E to AB and S to MT and CA.

Carex pyrenaica **Wahlenb. ssp.** ***micropoda*** **(C.A. Meyer) Hult.**
Pyrenean sedge
Habitat/Range: Moist to wet meadows and rocky slopes in the subalpine and alpine zones; frequent throughout BC except rare on the Queen Charlotte Islands and absent in NE BC; circumpolar, N to AK, YT and NT, E to SW AB and S to CO, UT and E OR; S Europe, E Asia.

Carex raynoldsii **Dewey**
Raynold's sedge
Habitat/Range: Moist to dry meadows and open forests in the steppe to subalpine zones; infrequent in S BC in and east of the Coast-Cascade Mountains; E to SW SK and S to CO, UT, NV, and CA.

Carex retrorsa **Schwein.**
Long-bracted or retrorse sedge
Habitat/Range: Bogs, swamps, shorelines, streambanks and wet meadows in the lowland, steppe and montane zones; frequent in S BC east of the Coast-Cascade Mountains and south of 56°N, rare in SW BC; N to NT, E to NF and S to NJ, OH, SD, CO, UT and NV.

Carex richardsonii **R. Br. in Richards.**
Richardson's sedge
Habitat/Range: Dry rocky slopes and forests in the montane zone; common in C and SC BC (Cariboo region); E to PQ and S to NY, IL, IN, OH and SD.

Carex rossii **W. Boott in Hook.** (*C. brevipes* W. Boott, *C. deflexa* Hornem. var. *rossii* [Boott in Hook.] Bailey)
Ross' sedge
Habitat/Range: Dry to mesic open forests, rock outcrops and forest openings in all vegetation zones; common throughout BC except absent on the Queen Charlotte Islands; N to AK and YT, E to MB, and S to MI, SD, CO and CA.
Notes: Part of the poorly studied *C. deflexa* complex.

***Carex rugosperma* Mack. var. *tonsa* [Fern.] Voss** (*C. tonsa* (Fern.) Bickn., *C. umbellata* Schkuhr ex. Willd. var. *tonsa* Fern.)

Bald sedge

Habitat/Range: Sand dunes in the montane zone; rare in EC BC, known only from Tête Jaune; E to PQ and NS, and S to IN, WI and MT.

***Carex rupestris* Bell. ex All.**

Curly or rock sedge

Habitat/Range: Dry rocky slopes, scree slopes and meadows in the alpine zone; rare in extreme N BC (ssp. *rupestris*) and rare in NW and SE BC (spp. *drummondiana*); ssp. *rupestris* - circumpolar, N to AK, YT and NT, and E to NF; Greenland, Iceland, Eurasia; spp. *drummondiana* - N to NT and S to CO and UT.

Notes: Two subspecies occur in BC.

1. Scales enveloping and largely concealing perigynia; spikes not slender, straight; leaves 2–3 mm wide; culms rather stout, stiff .... ssp. *drummondiana* (Dewey) Holub
1. Scales shorter than perigynia; spikes slender, flexuous; leaves about 1.5 mm wide; culms slender .... spp. *rupestris*

***Carex sartwellii* Dewey**

Sartwell's sedge

Habitat/Range: Marshes, fens, wet (often alkaline) meadows and moist open forests in the steppe and montane zones; infrequent in SC and SE BC; N to NT, E to ON and S to NY, IN, OH, PA, MO, and CO.

***Carex saxatilis* L. ssp. *laxa* [Trautv.] Kalela** (*C. physocarpa* Presl, *C. saxatilis* var. *major* Olney in S. Wats.)

Russet sedge

Habitat/Range: Bogs, fens, shorelines and wet meadows in the montane and subalpine zones; frequent throughout BC; circumpolar, N to AK, YT and NT, E to AB and S to CO, UT, NV and WA; N Eurasia.

***Carex saximontana* Mack.** (*C. backii* W. Boott in Hook. var. *saximontana* [Mack.] Boivin.)

Rocky Mountain sedge

Habitat/Range: Moist meadows in the montane and subalpine zones; infrequent in and E of the Coast-Cascade Mountains S of 55°N; E to PQ and NB and S to NY, UT and OR.

Notes: None of the material collected in BC represents *C. backii*, a species which is known from nearby WA. The following key distinguishes the two species.

1. Perigynia 5–6 mm long, upper third of the body empty; beaks stout, about 2 mm long, not serrulate; leaves deep green, flat .... *C. backii*
1. Perigynia 4–5 mm long, the upper part of the body filled by the achene; beaks conic, about 1 mm long, slightly serrulate; leaves glaucous-green; the margins revolute .. *C. saximontana*

***Carex scirpoidea* Michx.** (*C. pseudoscirpoidea* Rydb. = var. *pseudoscirpoidea*, *C. stenochlaena* Mack. = var. *stenochlaena*, *C. scirpiformis* Mack. = var. *scirpoidea*)

Single-spike sedge

Habitat/Range: Moist meadows and streambanks (var. *stenochlaena*) or mesic to dry meadows and open rocky slopes (vars. *pseudoscirpoidea* and var. *scirpoidea*) from the montane to alpine zones; var. *pseudoscirpoidea* - locally frequent in the Cascade Mountains, rare elsewhere in SC BC, var. *scirpoidea* - frequent in N and SE BC, var. *stenochlaena* - frequent in W BC, less frequent eastward; var. *pseudoscirpoidea* - S to CO, UT and CA, var. *scirpoidea* - amphiberingian, N to AK, YT and NT, E to NF and S to NH, MI, CO and WA; Greenland, E Asia, var. *stenochlaena* - N to AK and YT and S to MT, ID and E OR.

Notes: Three varieties occur in BC.

1. Perigynia relatively long and narrow, mostly 3–4.5 mm long and more than 2.5 times as long as wide; achenes distinctly slenderly stipitate; plants of moist sites mainly of W BC .... var. *stenochlaena* Holm

1. Perigynia relatively short and broad, mostly 2.5–3 mm long and less than 2.5 times as long as wide; achenes short-stipitate; plants of mesic to dry sites in SC BC or N and SE BC.

   2. Culms distinctly aphyllopodic, the flowering ones arising from vegetative shoots of the current year which have conspicuous basal scale leaves; plants mainly of N and SE BC .... var. *scirpoidea*

   2. Culms phyllopodic, clothed at base with the conspicuous dried-up leaves of the previous year; plants mainly in the Cascades .... var. *pseudoscirpoidea* (Rydb.) Cronq.

Carex scoparia **Schkuhr ex Willd.**
Pointed broom sedge
Habitat/Range: Moist to wet sites in the lowland and montane zone; rare in S BC; E to NF and S to AL, FL, NM and OR.

Carex scopulorum **Holm var. *bracteosa* [Bailey] F.J. Herm.** (*C. campylocarpa* Holm, *C. gymnoclada* Holm, *C. miserabilis* Mack; *C. scopulorum*)
Holm's Rocky Mountain sedge
Habitat/Range: Shorelines, streambanks and wet to moist meadows in the montane to alpine zones; rare in NW and SC BC, N to AK and YT, E to SW AB and S to CO, UT, NV and CA.

Carex simulata **Mack.**
Analogue sedge
Habitat/Range: Wet, often calcareous, fens in the montane zone; rare in SC BC (Chilcotin-Cariboo region); E to SK and S to CO, NM and CA.

Carex sitchensis **Prescott in Bong.** (*C. barbarae* Dewey; *C. aquatilis* Wahlenb. var. *dives* [Holm] Kukenth. var. *sitchensis* [Prescott in Bong.] Kelso nom. superfl.)
Sitka sedge
Habitat/Range: Swamps, shorelines, marshes, fens and wet meadows in the lowland and montane zones; common throughout BC south of 56°N; N to SE AK and S to N ID and CA.
Notes: Although Standley (1985) treated this taxon as a variety of *C. aquatilis*, we have decided, for practical reasons, to retain it at the species level. In BC it is often a community dominant and to have it merged into *C. aquatilis* sens. lat., by resource managers, is not desirable.

Carex spectabilis **Dewey** (*C. tolmiei* Boott in Hook.)
Showy sedge
Habitat/Range: Moist to mesic meadows and forest openings in the lowland to alpine zones; common in S BC, less frequent in C and NW BC; N to AK and YT, E to SW AB and S to CO, UT, NV and CA.
Notes: Sometimes difficult to separate from *C. podocarpa* in NW BC.

Carex sprengelii **Dewey ex Spreng.**
Sprengel's sedge
Habitat/Range: Moist to wet, open sites in the lowland and montane zones; rare in C, SC and SW BC; E to PQ and NB and S to NJ, NE and CO.

Carex stipata **Muhl. ex Willd.**
Awl-fruited or sawback sedge
Habitat/Range: Swamps, ditches, streambanks, lakeshores and wet meadows in the lowland, steppe and montane zones; frequent in BC S of 56°N, absent on the Queen Charlotte Islands; amphiberingian, N to SE AK, E to NF and S to FL, TX, NM and CA.

Carex stylosa **C.A. Meyer**
Long-styled sedge
Habitat/Range: Marshes, bogs and moist to wet meadows and gravelly sites in the lowland to alpine zones; frequent in coastal BC; circumpolar, N to AK, YT and NT, E to NF and S to NW WA; Greenland, N Europe and E Asia.

Carex supina **Willd. ex Wahlenb. ssp.** ***spaniocarpa*** **(Steud.) Hult.** (*C. spaniocarpa* Steud.)
Spreading arctic sedge
Habitat/Range: Dry, rocky, open sites in the montane zone; rare in N BC, infrequent in WC; circumpolar, N to AK, YT and NT, E to W ON, and S to MN; Greenland, Eurasia.

Carex sychnocephala **Carey**
Many-headed sedge
Habitat/Range: Shorelines and wet to moist meadows in the steppe and montane zones; rare in SC BC; N to S YT and S NT, E to S PQ and S to NY, SD, CO and N WA.

Carex tenera **Dewey**
Tender sedge
Habitat/Range: Mesic to dry meadows, shorelines and open forests in the steppe, montane and subalpine zones; rare in SC and SE BC; E to PQ, NB and NS and S to MO, NC, OH, OK and WY.

Carex tenuiflora **Wahl.**
Sparse-leaved sedge
Habitat/Range: Fens, bogs and moist forests in the montane zone; infrequent in BC east of the Coast Mountains and N of 52°N; circumpolar, N to AK, YT and NT, E to NF and S to MA, NY, MI and MN.

Carex tracyi **Mack.** (*C. leporina* auct. non L.)
Tracy's sedge
Habitat/Range: Moist to wet meadows in the lowland and montane zone; infrequent on Vancouver Island, rare elsewhere in BC; S to ID and CA.

Carex trisperma **Dewey**
Three-seeded sedge
Habitat/Range: Bogs and moist to wet forests in the montane zone; infrequent in BC east of the Coast-Cascade Mountains; E to NF and S to MD, IL and MN.

Carex unilateralis **Mack.**
One-sided sedge
Habitat/Range: Moist to wet sites in the lowland and montane zones; infrequent in SW BC; S to CA.

Carex utriculata **Boott in Hook.** (*C. rostrata non* Stokes in With., *C. rostrata* Stokes in With. var. *utriculata* [Boott in Hook.] Bailey)
Beaked or bottle sedge
Habitat/Range: Swamps, marshes, shorelines and wet meadows in the lowland, steppe and montane zones; common throughout BC except absent on the Queen Charlotte Islands; circumpolar, N to AK, YT and NT, E to NF and S to DE, WV, IN, SD, NM and CA; Greenland, Iceland, Eurasia.

Carex vaginata **Tausch**
Sheathed sedge
Habitat/Range: Swamps, bogs and wet meadows in the montane zone; frequent throughout BC east of the Coast-Cascade Mountains; circumpolar, N to AK, YT and NT, E to NF and S to NY, MI, MN, and UT.

***Carex viridula* Michx.** (*C. oederi* Retz. var. *viridula* [Michx.] Hult.)
Green sedge
Habitat/Range: Swamps, bogs, lakeshores and wet meadows in the lowland, steppe and montane zones; common in S BC, less frequent northward; amphiberingian, N to AK, YT and NT, E to NF and S to NJ, PA, ND, NM and CA; E Asia.

***Carex vulpinoidea* Michx.**
Fox sedge
Habitat/Range: Swamps, marshes, streambanks and wet meadows in the lowland, steppe and montane zones; rare in S BC; E to NF and S to FL, TX, AZ and OR.

***Carex xerantica* Bailey**
Dry-land sedge
Habitat/Range: Dry slopes and open forests in the steppe and montane zones; rare in SC and NE BC; E to S MB and S to MN, SD, CO and NM.

CYPERUS

1. Plants 0.3–2.0 dm tall; scales loose, the tips conspicuously recurved; spikelets borne in short, more or less capitate clusters .... *C. aristatus*
1. Plants 1–7 dm tall; scales closely set, the tips not recurved; spikelets borne in more or less elongate, cylindric spikes .... *C. erythrorhizos*

***Cyperus aristatus* Rottb.** (*C. inflexus* Muhl.)
Awned cyperus
Habitat/Range: Moist to wet, often sandy sites in the lowland and steppe zones; rare on S Vancouver Island and SC BC; cosmopolitan, E to PQ and NB and S to FL, TX and MX; Asia, S America, Africa.

***Cyperus erythrorhizos* Muhl.**
Red-rooted cyperus
Habitat/Range: Moist to wet lakeshores in the steppe zone; rare in SC BC, known only from Osoyoos Lake and Winfield; disjunct E to S ON and S to FL and CA.

DULICHIUM

***Dulichium arundinaceum* (L.) Britt.**
Three-way sedge
Habitat/Range: Wet meadows, lakeshores and streamsides in the lowland, steppe and montane zones; frequent in SW BC, rare in SC BC; E to NF and S to FL, TX and CA.

ELEOCHARIS

1. Stigmas 3; achenes 3-sided to nearly terete.
    2. Tubercles confluent with the achenes, merely conical, not forming distinct apical caps.
        3. Spikelets (5) 8–13 mm long, 10–20 flowered, fusiform; plants (2) 4–15 dm tall .. *E. rostellata*
        3. Spikelets 2–8 mm long, usually 2–9 flowered, ovoid; plants 0.2–3 (4) dm tall.
            4. Plants 0.2–0.6 (1) dm tall; lowest scale not subtending a flower; plants coastal .... *E. parvula*

4. Plants (0.5) 1–3 dm tall; lowest scale subtending a flower; plants not coastal .... *E. quinqueflora*

2. Tubercles forming distinct apical caps on the achenes.

5. Achenes white to pale gray or pale yellow, many-ribbed, the surfaces cross-ribbed; scales usually with a green or stramineous mid-vein, lowest one subtending a flower .... *E. acicularis*

5. Achenes yellow or golden, 3-ribbed, the surfaces finely cellular-roughened; scales dark with pale and hyaline margins, lowest one usually not subtending a flower .... *E. tenuis*

1. Stigmas 2; achenes lenticular (rarely 3-cleft in *E. ovata*).

6. Tubercles conspicuous, 1/4 to 1/2 the length of the achenes; plants rhizomatous perennials.

7. Tubercles rounded at the apices; spikelets with a single basal scale encircling the tops of the culms .... *E. kamtschatica*

7. Tubercles pyramidal; spikelets with 2 or 3 basal scales encircling the tops of the culms .... *E. palustris*

6. Tubercles flattened or a minute cone; plants annuals.

8. Achenes black to cherry red, 0.5–0.6 mm long, the bristles scarcely barbed and subequal to the achenes .... *E. atropurpurea*

8. Achenes stramineous to dark brown, (0.8) 1.0–1.5 mm long, the bristles conspicuously retrorsely barbed and equal to or exceeding the achenes .... *E. ovata*

Eleocharis acicularis **(L.) Roem. & Schult.**

Needle spike-rush

Habitat/Range: Marshes, mud flats and shorelines in the lowland, steppe and montane zones; frequent in S BC, infrequent northward; circumpolar, N to AK, YT and NT, E to NF and S to FL, N MX and CA; Greenland, Eurasia.

Eleocharis atropurpurea **(Retz.) Kunth**

Purple spike-rush

Habitat/Range: Shallow water and moist lakesides in the steppe zone; rare in SC BC, known only from Osoyoos Lake; pantropical, S to GA, NE, CO and CA; Eurasia, Africa, S America.

Eleocharis kamtschatica **(C.A. Meyer) Komarov**

Kamchatka spike-rush

Habitat/Range: Marshes, wet meadows and bog margins in the lowland zone; rare on N Vancouver Island, N Queen Charlotte Islands and adjacent N coast; amphiberingian, N to AK and scattered on the Canadian east coast; E Asia.

Eleocharis ovata **(Roth) Roem. & Schult.** (*E. obtusa* [Willd.] Schult. in Roem. & Schult.)

Blunt or ovoid spike-rush

Habitat/Range: Marshes, mud flats and wet meadows in the lowland and montane zones; frequent in SW BC, rare in the Queen Charlotte Islands, SC and SE BC; circumpolar, E to NF and S to CT, IN, MO and CA; Eurasia.

Eleocharis palustris **(L.) Roem. & Schult.** (*E. macrostachya* Britt. ex Small, *E. uniglumis* [Link] Schult. in Roem. & Schult.)

Common or creeping spike-rush

Habitat/Range: Wet meadows, ditches and shorelines in the lowland, steppe vegetation and montane zones; common throughout BC; circumpolar, N to AK, YT and NT, E to NF and S to NH, MD, ND, ID and OR; Greenland, Eurasia.

Notes: This is a highly variable polyploid complex which has yet to be studied adequately.

Eleocharis parvula **(Roem. & Schult.) Link ex Bluff & Fingerh.**
Small spike-rush
Habitat/Range: Salt marshes, mudflats and sandy shores along the coast; rare on S Vancouver Island and the lower mainland; circumpolar, S to N CA, Atlantic Coast; Europe.

Eleocharis quinqueflora **(Hartm.) Schwarz** (*E. pauciflora* [Lightf.] Link, *E. quinqueflora* ssp. *fernaldii* [Svens.] Hult. and ssp. *suksdorfiana* [Beauv.] Hult.)
Few-flowered spike-rush
Habitat/Range: Bog margins and wet (often calcareous) sites in the steppe and montane zones (rarely subalpine or alpine); infrequent in SC and SE BC, rare northward; circumpolar, N to AK, YT and NT, E to NF and S to NJ, PA, OH, IA, NM, AZ, N MX and CA; Greenland, Eurasia.

Eleocharis rostellata **(Torr.) Torr.**
Beaked spike-rush
Habitat/Range: Salt marshes, hot springs and alkaline or saline ponds in the lowland and steppe zones; rare in S BC; disjunct E to NS and throughout US; S America.

Eleocharis tenuis **(Willd.) Schult.** (*E. compressa* Sullivant, *E. elliptica* Kunth, *E. nitida* Fern. var. *borealis* [Svens.] Gleason)
Slender spike-rush
Habitat/Range: Lakeshores, streamsides and wet meadows in the steppe and montane zones; infrequent in SC and SE BC (N of 50°N) and EC and NE BC; N to SW AK, E to NF and S to GA, TX, MX and CO.

ERIOPHORUM

1. Spikelets, 2 or more, subtended by one or more leaf-like involucral bracts.
    2. Leaves narrow, elongate, channeled or triangular in cross-section; involucral bracts solitary, shorter than the inflorescence .... *E. gracile*
    2. Leaves mostly flattened, except at the apices; involucral bracts 2 or more with the longest equal to, or surpassing the inflorescence.
        3. Spikelets crowded, their scales with several conspicuous ribs; bristles orange-brown or coppery .... *E. virginicum*
        3. Spikelets loosely peduncled, their scales with a single rib; bristles white or dull yellow.
            4. Midrib of scales extending to the tip; sheaths green .... *E. viridicarinatum*
            4. Midrib of scales not extending to the tip; sheaths blackish .... *E. angustifolium*
1. Spikelets solitary, not subtended by leaf-like involucral bracts.
    5. Plants stoloniferous, usually with solitary culms.
        6. Bristles white; anthers 0.5–1 mm long; middle scales without pale margins ... *E. scheuchzeri*
        6. Bristles rust-colored; anthers 1–2.5 mm long; middle scales with pale margins .... *E. chamissonis*
    5. Plants without stolons, densely tufted.
        7. Scales with broad, pale margins, basal ones reflexed .... *E. vaginatum*
        7. Scales without pale margins, basal ones not reflexed.
            8. Bristles white; plants 0.6–2.5 dm tall .... *E. callitrix*
            8. Bristles cream-colored or yellow-tinged; plants 3–6 dm tall .... *E. brachyantherum*

Eriophorum angustifolium **Honck.** (*E. angustifolium* ssp. *scabriusculum* Hult., ssp. *subarcticum* [Vassill.] Hult. and ssp. *triste* [Fries] Hult., *E. polystachion* L. *pro parte*)
Narrow-leaved cotton-grass
Habitat/Range: Bogs, marshes, shorelines and wet meadows in all vegetation zones; common throughout BC, circumpolar, N to AK, YT and NT, E to NF and S to NH, IL, MI, CO, NM and OR; Eurasia.

Eriophorum brachyantherum **Trautv. & Meyer**
Short-anthered cotton-grass
Habitat/Range: Bogs and wet (often calcareous) meadows in the montane to alpine zones; common in BC north of 52°N and in and E of the Coast-Cascade Mountains; circumboreal, N to AK, YT and NT, and E to NF; Eurasia.

Eriophorum callitrix **Cham.**
Arctic cotton-grass
Habitat/Range: Bogs and wet to mesic (often calcareous) meadows in the montane to alpine zones; infrequent in BC north of 57°N and E of the Coast-Cascade Mountains; amphiberingian, N to AK, YT and NT and E to NF; Greenland, E Asia.

Eriophorum chamissonis **C.A. Meyer** (*E. altaicum* Meinsh. var. *neogaeum* Raymond, *E. russeolum* Fries in Hartm.)
Chamisso's cotton-grass
Habitat/Range: Bogs, streambanks, lakeshores and wet meadows in all vegetation zones; common throughout BC; N to AK, YT and NT, E to NF and S to MN, CO, UT and OR.

Eriophorum gracile **Koch in Roth**
Slender cotton-grass
Habitat/Range: Bogs, marshes and lakeshores in the lowland, steppe and montane zones; infrequent in BC, absent from NW BC, Queen Charlotte Islands and Vancouver Island; circumpolar, N to AK and NT, E to NF and S to MN, NE, CO and N CA; Eurasia.

Eriophorum scheuchzeri **Hoppe**
Scheuchzer's cotton-grass
Habitat/Range: Marshes, lakeshores and wet meadows in the montane to alpine zones; frequent in N BC, rare in S BC east of the Coast-Cascade Mountains; N to AK, YT and NT, E to NF and S to CO; Eurasia.

Eriophorum vaginatum **L. ssp. *spissum*** **(Fern.) Hult.**
Sheathed cotton-grass
Habitat/Range: Bogs, marshes and wet meadows in the montane to alpine zones; rare in NC and NE BC; circumpolar, N to AK, YT and NT and E to NF; Eurasia.

Eriophorum viridicarinatum **(Engelm.) Fern.** (*E. polystachion* L. var. *viridicarinatum* Engelm.)
Green-keeled cotton-grass
Habitat/Range: Bogs, marshes and wet meadows in the montane to alpine zones; frequent in NW and E BC, infrequent in SC BC; N to AK and YT, E to NF and S to NY, IA, OH, MI, CO and ID.
Notes: Very similar to *E. polystachion* and possibly best treated as a variety of the latter.

Eriophorum virginicum **L.**
Tawny cotton-grass
Habitat/Range: Bogs and wet ditches in the lowland zone; rare in SW BC, known only from the Fraser River delta; introduced from E North America.

HEMICARPHA

***Hemicarpha micrantha* (Vahl) Britt.** (*Lipocarpha micrantha* [Vahl] G. Tucker)
Small-flowered hemicarpha
Habitat/Range: Lakeshores in the steppe zone; rare in SC BC, known only from Osoyoos Lake; disjunct E to ON and S to FL, MN, WA, ID and CA.

KOBRESIA

1. Spikelets arranged in spikelike panicles, the spikes 4–11 mm wide .... *K. simpliciuscula*
1. Spikelets arranged in compact spikes, the spikes 2–7 mm wide.
   2. Scales subtending spikelets 4.5–6 mm long; spikes 4–7 mm wide .... *K. sibirica*
   2. Scales subtending spikelets 2.5–4.5 mm long; spikes 2–3 mm wide .... *K. myosuroides*

***Kobresia myosuroides* (Vill.) Fiori & Paol.** (*K. bellardii* [All.] Degl. ex Loise)
Bellard's kobresia
Habitat/Range: Mesic to dry meadows and rock outcrops in the montane to alpine zones; frequent in BC in and E of the Coast-Cascade Mountains; circumpolar, N to AK, YT and NT, E to Labr. and S to CO, UT and CA; Eurasia.

***Kobresia sibirica* Turcz.**
Siberian kobresia
Habitat/Range: Bogs and wet meadows in the upper montane to alpine zone; rare in N BC, known only from the Liard Plateau; amphiberingian, N to AK, YT and NT; E Asia.

***Kobresia simpliciuscula* (Wahl.) Mack.**
Simple kobresia
Habitat/Range: Bogs, wet meadows and shallow ponds in the montane and subalpine zones; infrequent in extreme N BC; circumpolar, N to AK, YT and NT, E to NF and S to ID, CO and OR; Eurasia.

RHYNCHOSPORA

1. Spikelets white, becoming pale-brown; achenes 1.5–2.5 mm long, tubercles 1/3 to 1/2 as long as the achenes; plants frequent in and W of the Coast-Cascade Mountains .... *R. alba*
1. Spikelets reddish-brown to dark brown; achenes 2.5–4.2 mm long; tubercles almost as long as the achenes; plants rare in SC BC .... *R. capillacea*

***Rhynchospora alba* (L.) Vahl**
White beak-rush
Habitat/Range: Bogs and fens in the lowland and montane zones; frequent in and W of the Coast-Cascade Mountains in BC south of 56°N; circumpolar, N to SW AK, disjunctly E to NF and S to NC, OH, MN, ID and CA; West Indies, Eurasia.

***Rhynchospora capillacea* Torr.**
Brown beak-rush
Habitat/Range: Calcareous bogs, swamps and shorelines in the steppe zone; rare in SC BC, known only from the Salmon Arm area; E to NF and S to TN, VA, OH, IN and SD.

SCIRPUS

1. Plants annuals or perennials; bristles absent; achenes scarcely apiculate.
   2. Involucral bracts more than 2 cm long .... *S. supinus*
   2. Involucral bracts less than 2 cm long.
      3. Involucral bracts subequal to slightly surpassing the solitary spikelet; achenes smooth, reddish-brown, dull .... *S. cernuus*
      3. Involucral bracts far surpassing the cluster of 2–3 spikelets; achenes longitudinally ribbed, dark brown, shiny .... *S. setaceus*
1. Plants perennials; bristles present or achenes apiculate, or both.
   4. Involucral bracts 2 or more, leaf-like and more or less spreading.
      5. Spikelets 1.2–2.5 cm long; scales puberulent.
         6. Stigmas 2; bristles poorly developed, up to about half as long as the achenes, or obsolete; rhizomes thickened at the nodes, these with numerous conspicuous secondary rootlets .... *S. maritimus*
         6. Stigmas 3; bristles well developed, usually as long as the achenes; rhizomes bearing firm tubers, these with a few inconspicuous secondary rootlets .... *S. fluviatilis*
      5. Spikelets 0.3–0.8 cm long; scales glabrous.
         7. Bristles not barbed, long, flexuous, far surpassing the achenes and scales; stigmas 3 .... *S. cyperinus*
         7. Bristles retrorsely barbed, subequal to slightly surpassing the achenes but not the scales; stigmas 2 or 3.
            8. Sheaths red-tinged, at least basal ones; bristles 4, barbed nearly to the base; stigmas 2 .... *S. microcarpus*
            8. Sheaths not red-tinged, bristles usually 6, barbed only above the middle; stigmas 3 .... *S. pallidus*
   4. Involucral bract solitary, not leaf-like and spreading (except in *S. nevadensis*), appearing as an extension of the culm.
      9. Spikelets more or less numerous in branching inflorescences, often sessile in clusters at the ends of branches.
         10. Spikelets reddish-brown; scales about equal to the achenes, essentially glabrous except for the pubescent margins .... *S. validus*
         10. Spikelets greyish-brown; scales surpassing the achenes, more or less viscid-puberulent .... *S. acutus*
      9. Spikelets solitary or few, sessile in a sessile cluster.
         11. Spikelets solitary (rarely 2); plants usually aquatic; the flaccid stems usually floating distally rather than emersed .... *S. subterminalis*
         11. Spikelets 2 or more; plants sometimes growing in water, but stems stiff and emergent for most of their length.
            12. Achenes evidently cellular-reticulate (at 10X), culms subterete ...... *S. nevadensis*
            12. Achenes not cellular-reticulate, culms triquetrous.

13. Involucral bracts solitary; culms sharply triquetrous with concave sides; achenes 1.8–2.5 mm long .... *S. olneyi*

13. Involucral bracts 2 or 3, the second and third ones resembling enlarged scales of the spikelet; culms not sharply triquetrous, the sides only slight concave or slightly convex; achenes 2.2–3.3 mm .... *S. americanus*

***Scirpus acutus* Muhl. in Bigel.** (*S. lacustris* L. ssp. *glaucus* [Reichenb.] Hartm., *S. occidentalis* [S. Wats.] Chase, *S. tabernaemontanii* Gmel.)
Hand-stemmed bulrush
Habitat/Range: Marshes, lakeshores and streams in the lowland, steppe and montane zones; common throughout BC south of 56°N, less frequent northward; circumpolar, N to AK and YT, E to NF and S to TN, GA, TX, NM and CA; Eurasia, C. America.

***Scirpus americanus* Pers.** (*S. longispicatus* [Britt.] Smyth, *S. monophyllus* Presl, *S. pungens* Vahl)
American bulrush
Habitat/Range: Marshes, lakeshores, saline shores and wet meadows in the lowland and steppe zones; common in SW and SC BC, rare in SE BC; cosmopolitan, disjunct N to EC AK, E to NF and S to FL, TX, MX and CA; S America, S Australia, New Zealand.
Notes: The infraspecific taxonomy in this species has yet to be elucidated and certainly is not clear in BC.

***Scirpus cernuus* Vahl** (*S. cernuus* var. *californicus* [Torr.] Beetle)
Low clubrush
Habitat/Range: Marshes, tidal flats and beaches in the lowland zone; common in SW BC, rare northward along the coast; S to CA and NW MX; cosmopolitan, Eurasia, S America, Australia and New Zealand.

***Scirpus cyperinus* (L.) Kunth var. *brachypodus* (Fern.) Gilly** (*S. atrocinctus* Fern.)
Wool-grass
Habitat/Range: Swamps and wet meadows in the lowland and montane zones; infrequent in extreme SW BC, rare in SE BC and the Queen Charlotte Islands; E to NF and S to FL, NC, OK, TX and WA.

***Scirpus fluviatilis* (Torr.) Gray**
River bulrush
Habitat/Range: Marshes, streamsides and lakeshores in the lowland and steppe zone; rare in SW and SC BC; amphiberingian, E to PQ and NB and S to VA, KS, NM and CA; E Asia.
Notes: Since this species may often lack fruits the rhizome becomes a critical feature for identification.

***Scirpus maritimus* L. var. *paludosus* (A. Nels.) Kuekenth.** (*S. paludosus* A. Nels., *S. robustus* auct. non Pursh)
Seacoast or alkali bulrush
Habitat/Range: Marshes, swamps, saline and alkaline ponds and wet meadows in the lowland and steppe zones; frequent in BC south of 53°N; N to AK, YT and NT, E to NS and S to OK, MO, N MX and CA.

***Scirpus microcarpus* Presl** (*S. sylvaticus* auct. non L., *S. sylvaticus* ssp. *digynus* [Boeckl.] Koyama)
Small-flowered bulrush
Habitat/Range: Marshes, swamps and wet meadows in the lowland, steppe and montane zones; common throughout BC except absent in NW BC; amphiberingian, N to AK, E to NF and S to VA, MI, NE, NM and CA; E Asia.

Scirpus nevadensis **S. Wats.**
Nevada bulrush or clubrush
Habitat/Range: Wet, usually alkaline, meadows in the lowland, steppe and montane zones; infrequent in SW and SC BC; E to MB and S to ND, UT and CA; S America.

Scirpus olneyi **A. Gray**
Olney's bulrush
Habitat/Range: Saline or alkaline marshes and wet meadows in the lowland and montane zones; rare in SW BC, known only from Qualicum Beach and Pemberton; disjunctly E to NS and S to NH, OH, MO, FL, MX and CA; S America.

Scirpus pallidus **(Britt.) Fern.** (*S. atrovirens* Willd. var. *pallidus* Britt.)
Pale bulrush
Habitat/Range: Swamps and wet meadows in the lowland and montane zones; rare in S BC; E to ON and S to MN, MO, TX, UT, NM and NE OR.

Scirpus setaceus **L.** (*Isolepis setaceus* [L.] R. Br.)
Bristle clubrush
Habitat/Range: Disturbed sites in the lowland zone; rare in SW BC, known only from Victoria; introduced from Eurasia.

Scirpus subterminalis **Torr.**
Water clubrush
Habitat/Range: Shallow ponds and streams in the lowland zone; infrequent in coastal and SC BC; N to AK, disjunctly E to NF and S to GA, IN, MO, MT and OR.

Scirpus supinus **L var. *saximontanus* (Fern.) Koyama** (*S. saximontanus* Fern.)
European clubrush
Habitat/Range: Shallow water in the montane zone; rare in SC BC, known only from Shuswap Lake; known elsewhere only from Europe.

Scirpus validus **Vahl** (*S. lacustris* L. ssp. *validus* [Vahl] Koyama)
Soft-stemmed bulrush
Habitat/Range: Marshes, lakeshores and streams in the lowland, steppe and montane zones; common throughout BC; N to AK and YT, E to NF and S to TN, GA, MO, TX, NM and CA; C. America

TRICHOPHORUM

1. Culms sharply 3-angled, scabrous above; bristles extending 1–2.5 cm beyond the spikelets .... *T. alpinum*
1. Culms cylindrical, smooth; bristles scarcely, if at all, surpassing the spikelets or absent.
    2. Culms rhizomatous; scales awnless or nearly so; spikelets ovate; bristles absent ...... *T. pumilum*
    2. Culms cespitose; scales, at least the lower ones, awned; spikelets lanceolate; bristles present ... . *T. cespitosum*

Trichophorum alpinum **(L.) Pers.** (*Scirpus hudsonianus* [Michx.] Fern.)
Hudson Bay clubrush
Habitat/Range: Bogs, swamps and wet meadows in the montane and subalpine zones; infrequent in BC east of the Coast Mountains; circumpolar; N to AK, YT and NT, E to NF and S to NH and MN; Eurasia.

***Trichophorum cespitosum* (L.) Hartm.** (*Scirpus cespitosus* L.)
Tufted clubrush
Habitat/Range: Bogs, marshes and wet meadows in all vegetation zones; common throughout BC; circumboreal, N to AK, YT and NT, E to NF and S to NC, TN, MN, UT and OR; Greenland, Eurasia.

***Trichophorum pumilum* (Vahl.) Schinz & Thell.** (*Scirpus pumilus* Vahl, S. *rollandii* Fern.)
Dwarf clubrush
Habitat/Range: Bogs, lakeshores and wet meadows in the subalpine and alpine zones; rare in NE and EC BC, known only from Muncho Lake and McBride; circumpolar, N to YT and NT, E to PQ and S to CO; Eurasia.

HYDROCHARITACEAE

1. Leaves few, alternate, basal and ribbon-like .... *Vallisneria*
1. Leaves numerous, opposite or whorled on elongate stems, not ribbon-like.
   2. Main leaves usually 2.5 cm long, or longer, in whorls of 3–6 .... *Egeria*
   2. Main leaves less than 2.5 cm long, opposite or in whorls of 3 .... *Elodea*

EGERIA

***Egeria densa* Planch.** (*Elodea densa* [Planch.] Casp.)
Brazilian waterweed
Habitat/Range: Lakes, ponds and streams; rare in SW BC, known only from the Victoria area; introduced from S America.

ELODEA

1. Leaves broadly lanceolate to oblong or ovate, usually obtuse .... *E. canadensis*
1. Leaves linear or narrowly linear - lanceolate, acuminate .... *E. nuttallii*

***Elodea canadensis* Rich. in Michx.** (*Anacharis canadensis* [Rich. in Michx.] Planch.)
Canadian waterweed
Habitat/Range: Lake, ponds, streams and ditches in the lowland, steppe and montane zones; common in extreme S BC, rare northward; E to PQ and NS and S to AR, NC, UT, NM, AZ and CA.

***Elodea nuttallii* (Planch.) St. John** (*Anacharis nuttallii* Planch.)
Nuttall's waterweed
Habitat/Range: Lakes, ponds and streams in the lowland, steppe and montane zones; rare in S BC; E to ON and SW PQ and S to NC, AL, MS, MO, CO and CA.

VALLISNERIA

***Vallisneria americana* Michx.**
American tapegrass
Habitat/Range: Lakes, ponds and streams in the lowland and montane zones; rare in SW and SC BC; introduced from E North America.

IRIDACEAE[6]

1. Flowers orange-yellow, asymmetrical, curved to one side; plants from corms as well as rhizomes; rare garden escapes .... *Crocosmia*
1. Flowers blue, purple, white or bright clear yellow, symmetrical; plants lacking corms.
   2. Perianth segments of two distinct types; flowers large, greater than 5 cm wide .... *Iris*
   2. Perianth segments all alike; flowers mostly small, less than 5 cm wide.
      3. Perianth segments pale to deep reddish-purple (rarely white), 15–25 mm long; filaments connate one-third to one-half their length; flowers usually strongly nodding .... *Olsynium*
      3. Perianth segments pale to deep blue or yellow (rarely white), 5–15 (20) mm long; filaments connate to near their tips; flowers usually held erect or horizontally .... *Sisyrinchium*

CROCOSMIA

***Crocosmia x crocosmiiflora* (V. Lem. ex E. Morr.) N.E. Br.**
Montbretia
Habitat/Range: Waste places and edges of forests in the lowland zone; rare garden escape on Vancouver Island and the lower mainland; introduced from South Africa.

IRIS

1. Flowers yellow; plants usually more than 1 m tall .... *I. pseudacorus*
1. Flowers predominately blue, purple or white; plants usually less than 1 m tall.
   2. Principal leaves more than 1 cm wide; petals (standards) 0.5–2 cm long, sharp pointed .... *I. setosa*[7]
   2. Principal leaves less than 1 cm wide; petals (standards) 3–6 cm long, blunt-tipped.
      3. Blades of sepals (falls) oblanceolate; bracts green (or reddish) and not papery at flowering time .... *I. missouriensis*
      3. Blades of sepals (falls) broadly rounded; bracts brown and papery at flowering time .... *I. sibirica*

---

6 Several species and hybrids in the genus *Crocus* are cultivated commonly and often persist around old homesites, in waste places and along roadsides, but they are not known to be truly naturalized in our flora. They are small scapose, cormous plants, with narrow grass-like leaves, with a white stripe along the midrib, and relatively large white, yellow or purple flowers in early spring.

7 Often confused with *Iris missouriensis* and all material from BC identified as *I. setosa* seems to be the former. As *I. setosa* has been found in adjacent parts of Alaska and the Yukon and may yet be found in NW BC, it has been retained in the key.

***Iris missouriensis* Nutt.**

Western blue iris

Habitat/Range: Moist sites in edges of lakes and marshy meadows in the lowland and montane zones; rare in W BC, known from scattered localities, some or all of which may be introduced; E to SE AB and S to ND, SD, CA and N MX.

***Iris pseudacorus* L.**

Yellow iris

Habitat/Range: Moist sites in ditches, sloughs and streambanks in the lowland and steppe zones; frequent garden escape in S BC; introduced from Europe.

***Iris sibirica* L.**

Siberian iris

Habitat/Range: Waste places, roadsides and abandoned gardens in the lowland zone; rare garden escape in S BC; introduced from Asia.

OLSYNIUM

***Olsynium douglasii* (A. Dietr.) Bickn.** (*Sisyrinchium douglasii* A. Dietr.; *S. inflatum* [Suksd.] St. John = var. *inflatum*)

Satinflower or Douglas' blue-eyed-grass (var. *douglasii*) and purple blue-eyed-grass (var. *inflatum*)

Habitat/Range: Dry rocky bluffs, oak woodlands and sagebrush slopes in the lowland and steppe zones; var. *douglasii* locally frequent on S Vancouver Island and the Gulf Islands; var. *inflatum* rare in SC BC; S to ID, N UT and N CA.

Notes: Two varieties occur in BC.

1. Perianth segments dark red-purple; filament tubes only slightly enlarged above the base; flowers commonly 2 per stem; plants of the southern coast .... var. *douglasii*

1. Perianth segments pale purple; filament tubes with an inflated area just above the base; flowers commonly 3 per stem; plants of the dry southern interior .... var. *inflatum* (Suksd.) Cholewa & Henderson

SISYRINCHIUM [8]

1. Flowers yellow; capsules oblong; plants becoming black on drying .... *S. californicum*
1. Flowers blue; capsules round or ovoid; plants becoming dull, but not black on drying.
    2. Outer bracts of spathes nearly twice the length of the inner, or longer, generally united basally for no more than 3.5 mm; stems generally broader than the leaves.
        3. Perianth segments elliptic, often deeply retuse or emarginate and aristate; outer bracts often widened above the apex of the inner; inner bracts often abruptly protruding basally from the outer, giving the spathes a somewhat gibbous appearance; pedicels generally 19–30 mm long, long, often exceeding the inner bracts .... *S. montanum*
        3. Perianth segments oblanceolate, the apex rounded and aristate; outer spathe bracts tapering uniformly to a slender, acute apex; inner bracts not gibbous; pedicels generally 14–18 mm long, often shorter than the inner bracts .... *S. septentrionale*
    2. Outer bracts of spathes usually much less than twice the length of the inner, united for 4–7 mm; stems generally equal to or narrower than the leaves.

---

8 A difficult genus to which many names have been applied. Key and taxonomic treatment modified after Henderson (1976).

4. Leaves 2.5–5.5 mm wide (averaging 3.8 mm); stems 1.7–3.5 mm wide (averaging 2.4 mm); outer bracts of spathes united for about 5 mm; styles generally not exserted beyond the anthers .... *S. littorale*

4. Leaves and stems narrower, leaves averaging less than 3.8 mm, stems averaging less than 2.4 mm; outer bract of spathes united for about 4 mm; styles generally exserted beyond the anthers .... *S. idahoense*

Sisyrinchium californicum **(Ker-Gawl.) Dryand. in Ait.**
Golden-eyed-grass
Habitat/Range: Moist edges of ponds, seepage areas and bogs in the lowland zone; infrequent on S and W Vancouver Island and the Gulf Islands; S to S CA.

Sisyrinchium idahoense **Bickn.** (*S. angustifolium* auct. non Mill.; *S. macounii* Bickn. = var. *macounii*)
Idaho blue-eyed-grass
Habitat/Range: Moist grassy meadows in the lowland, steppe and montane zones; var. *idahoense* infrequent throughout S BC, var. *macounii* rare on S Vancouver Island and the Gulf Islands; S to MT, ID and N CA.
Notes: Two varieties occur in BC.

1. Stems narrow, generally averaging less than 1.5 mm wide; perianth segments usually more than 15 mm long .... var. *macounii* (Bickn.) Henderson

1. Stems wider, generally averaging more than 1.5 mm wide; perianth segments usually less than 15 mm long .... var. *idahoense*

Sisyrinchium littorale **Greene**
Shore blue-eyed-grass
Habitat/Range: Coastal meadows in the lowland zone; infrequent in coastal BC; N to AK and S to WA.

Sisyrinchium montanum **Greene** (*S. angustifolium* auct. non Mill.)
Mountain blue-eyed-grass
Habitat/Range: Moist meadows and streambanks in the steppe and montane zones; frequent throughout BC east of the Coast-Cascade Mountains; E to AB and S to CO, WY and MT.

Sisyrinchium septentrionale **Bickn.** (*S. angustifolium* auct. non Mill.)
Northern blue-eyed-grass
Habitat/Range: Dry meadows and streambanks in the steppe and montane zones; infrequent in BC east of the Coast-Cascade Mountains, E to MB and S to WA.

JUNCACEAE[9]

1. Fruits 3-chambered; seeds many; leaves stiff, glabrous; leaf sheaths open .... *Juncus*

1. Fruits 1-chambered; seeds 3; leaves flexible, usually hairy on margins at base; leaf sheaths closed .... *Luzula*

JUNCUS

1. Involucral bract resembling an extension and continuation of the stem, the inflorescence appearing lateral.

    2. Involucral bract channeled along the upper side .... *J. vaseyi*

---

9 Contributed by A. Ceska.

2. Involucral bract cylindrical, not channelled.
    3. Flowers usually only 2–3 (5); seeds with tail-like appendages on each end (caudate).
        4. Uppermost of the basal sheaths of the stem with a well-developed blade, 2–7 cm long; capsule acute .... *J. parryi*
        4. Uppermost of the basal sheaths of the stem bladeless, or with a blade less than 1 cm long; capsule retuse .... *J. drummondii*
    3. Flowers several or numerous; seeds cylindrical without tail-like appendages.
        5. Stems in dense clusters.
            6. Stem and lowest bract coarsely sulcate with 10–20 longitudinal ridges, epidermal cell on the ridges distinctly larger than the others .... *J. conglomeratus*
            6. Stem and lowest bract smooth to finely striate with 30–60 longitudinal ridges, epidermal cells more or less uniform .... *J. effusus*
        5. Stems distributed along elongated rhizomes, not in dense clusters.
            7. Perianth 6–7 mm long; stems flattened, twisted .... *J. breweri*
            7. Perianth less than 6 mm long; stems round or filiform, usually straight.
                8. Stem finely grooved; sheaths dull, light-brown; inflorescence in the middle of the 'stem'; perianth whitish .... *J. filiformis*
                8. Stem smooth; sheath shiny brown; inflorescence above the middle, in the upper third of the 'stem'; perianth red brown.
                    9. Anthers longer or equal to filaments; flowers 8–20; inflorescence lax .... *J. balticus*
                    9. Anthers shorter than filaments; flowers 3–12; inflorescence dense .... *J. arcticus*

1. Involucral bract not forming continuation of the stem, inflorescence ± terminal.
    10. Plants annual.
        11. Flowers solitary on the scape, rarely in groups of two .... *J. kelloggii*
        11. Flowers in long and diffuse inflorescence, more than one flower on a scape ..... *J. bufonius*
    10. Plants perennial.
        12. Flowers each with 2 involucral bracteoles, usually either single or slightly apart, forming an open inflorescence.
            13. Stem with several leaves; outer perianth segments obtuse with incurved or hooded tips .... *J. gerardii*
            13. Stem leafless or with 1 leaf; outer perianth segments acute with ascending or spreading tips.
                14. Seeds with long appendages; capsules longer than the perianth .... *J. vaseyi*
                14. Seeds lacking appendages; capsules shorter than or as long as the perianth .... *J. tenuis*
        12. Flowers without bracteoles, clustered into small groups or heads; these heads being either single, or (more often) forming a compound inflorescence.
            15. Leaf blades not at all septate, or with septa just close to their tips.

16. Stems without leaves, all leaves basal; plants less than 20 cm tall.

17. Capsules distinctly retuse; inflorescence usually 2-flowered .... *J. biglumis*

17. Capsules obtuse to acute; inflorescence 3–5 flowered.

18. Bracts long obtuse, lower mucronate, smaller than flowers; capsules dark, larger than perianth .... *J. triglumis*

18. Bracts long acuminate or awned, larger than flowers; capsules pale, smaller than or equal to perianth .... *J. albescens*

16. Leaves along the stem; plants often more than 25 cm tall.

19. Capsules about 1.5-times longer than perianth.

20. Capsules 6–8 mm long, chestnut brown; anthers 1.0–1.5 mm long; plants with stolons .... *J. castaneus*

20. Capsules 5–6 mm long, pale brown; anthers 0.4–0.7 mm long; plants without stolons .... *J. stygius*

19. Capsules not longer or only slightly longer than the perianth.

21. Heads usually hemispheric; perianth deep brown with lighter middle part; seeds with appendages .... *J. regelii*

21. Heads not hemispheric; perianth pale brown, largely scarious; seeds without appendages.

22. Capsules shorter than the perianth; leaves cauline, soft .... *J. longistylis*

22. Capsules equal or slightly longer than the perianth; leaves mostly basal, stiff.

23. Seeds 0.5 mm long, wedge-like, angular, broadest at the tip and narrowing down to the base .... *J. falcatus*

23. Seeds 0.3 mm long, cylindrical, broadest in the middle .... *J. covillei*

15. Leaf blades more or less definitely septate.

24. Blades strongly flattened, the septae not complete, running from the edges to about 2/3 of the leaf width.

25. Capsules obtuse; perianth blunt, mostly deep brown; heads fewer than 10 .... *J. ensifolius*

25. Capsules acute; perianth acuminate, greenish yellow to pale brown; heads more than 10 .... *J. oxymeris*

24. Blades rounded or little flattened, the septae complete across the full width of the leaf.

26. Early leaves filiform, flaccid, usually submerged; stem decumbent, rooting at the nodes; inflorescence often viviparous, with young small plants.

27. Perianth 3–4.5 mm long, narrowly lanceolate, acute or acuminate; capsules acute, black, 1–1.5 mm longer than perianth .... *J. supiniformis*

27. Perianth 2.5–3 mm long, obtuse or outer acute; capsules obtuse or retuse, brown, less than 1 mm longer than the perianth ... *J. bulbosus*

26. Early leaves stiff, not filiform; plants erect or decumbent, but not rooting in the nodes; inflorescence not viviparous.

28. Heads many-flowered, spheroidal or globose.

29. Plants with a single terminal head .... *J. mertensianus*

29. Plants with several heads.

30. Capsules ovoid-cylindric, abruptly narrowed or rounded, not longer than perianth.

31. Stamens 3, anthers shorter than filaments; perianth narrowly acuminate .... *J. bolanderi*

31. Stamens 6, anthers longer than filaments; perianth lanceolate to acuminate .... *J. nevadensis*

30. Capsules uniformly tapered into a short tip, longer than the perianth.

32. Perianth up to 3 mm long, deep brown; rhizomes without nodes .... *J. articulatus*

32. Perianth longer than 3 mm, greenish brown to tawny; rhizomes with swollen nodes.

33. Inner perianth segments shorter than the outer, 4–5.5 mm long; heads 1–1.5 cm when pressed .. .. *J. torreyi*

33. Inner perianth segments longer than the outer ones, 3-4 mm long; heads less than 1 cm when pressed .... *J. nodosus*

28. Heads with few flowers, obconical.

34. Capsules twice as long as perianth, over 5 mm long .... *J. supiniformis*

34. Capsules less than twice as long as the perianth, shorter than 5 mm.

35. Perianth segments blunt and rounded; capsule rounded .... *J. alpinoarticulatus*

35. Perianth segments acute or acuminate; capsule tapered to tip.

36. Capsules longer than perianth; perianth (2) 2.5–3 mm long; stamens 6 .... *J. articulatus*

36. Capsules about equal to perianth; perianth 3 mm long or longer; stamens 3.

37. Inner perianth segments equal to the outer ones; seeds without appendages .... *J. acuminatus*

37. Inner perianth segments longer than outer ones; seeds with tail-like appendages .... *J. brevicaudatus*

***Juncus acuminatus* Michx.**

Tapered rush

Habitat/Range: Wet lakeshores, marshes and bogs in the lowland and montane zones; frequent in SW BC, infrequent E of the Coast-Cascade Mountains; disjunct E to ON, PQ and NS, and S to ME, MA, MI, WI, MN, WA and CA.

***Juncus albescens* (Lange) Fern.** (*J. triglumis* L. ssp. *albescens* [Lange] Hult. and var. *albescens* Lange)

Whitish rush

Habitat/Range: Wet calcareous fens in the montane zone; rare in C BC (Chilcotin Plateau); N to AK, YT and NW NT, S to WY and CO.

***Juncus alpinoarticulatus* Chaix in Vill.** (*J. alpinus* Vill., *J. geniculatus* Schrank)

Alpine rush

Habitat/Range: Lakeshores, riverbanks, bogs and marshes from the lowland to subalpine zones; infrequent throughout BC; circumpolar, N to AK, YT and NT, E to NF and S to VT, NY, IL, MN, NE, and WA; Greenland, Eurasia.

***Juncus arcticus* Willd.** (*J. haenkei* E. Meyer = ssp. *sitchensis*, *Juncus balticus* var. *alaskanus* [Hult.] Pors.)

Arctic rush

Habitat/Range: Tidal flats and lake margins in the lowland and montane zones; common in coastal BC; circumpolar, N to AK, YT and NT, E to NF and S to CA; Eurasia.

Notes: The taxonomic relationship between this species and related *J. balticus* has to be clarified. *J. arcticus* is a variable species, with many infraspecific taxa described. The whole complex requires further study and more intensive collecting throughout our area. We treat all our plants as ssp. *sitchensis* with the exception of some plants from northern BC which belong to the rare ssp. *alaskanus*.

1. Involucral bract 10–30 cm long; inner perianth segments acute, narrowly scarious margined .... ssp. *sitchensis* Engelm.

1. Involucral bract 2–5 cm long; inner perianth leaves short-acuminate, broadly scarious-margined .... ssp. *alaskanus* Hult.

***Juncus articulatus* L.**

Jointed rush

Habitat/Range: Wet lakeshores, riverbanks and disturbed sites in the lowland, steppe and montane zones; common in BC south of 55°N; circumpolar, N to AK, E to NF and S to NJ, WV, IL, MN, and OR; Eurasia.

***Juncus balticus* Willd.** (*J. arcticus* Willd. ssp. *ater* [Rydb.] Hult., *J. arcticus* ssp. *balticus* [Willd.] Hyl., *J. balticus* var. *littoralis* Engelm., *J. vallicola* Rydb.)

Baltic rush

Habitat/Range: Wet lakeshores and sandy river bars in the lowland, steppe and montane zones; infrequent throughout BC except locally common in the Cariboo-Chilcotin region; circumpolar, N to AK, YT and NT, E to PQ and NF, and S to PA, IN, IL, CO, and CA; Eurasia.

***Juncus biglumis* L.**

Two-flowered rush

Habitat/Range: Wet seepy places, mossy ledges and marshes in the subalpine and alpine zones; infrequent in BC north of 55°N; circumpolar, N to AK, YT and NT, and E to N MB, ON and PQ; Greenland, Eurasia.

***Juncus bolanderi* Engelm.**

Bolander's rush

Habitat/Range: Wet meadows and lake margins in the lowland zone; rare in SW BC; S to CA.

***Juncus brevicaudatus* (Engelm.) Fern.**
Short-tailed rush
Habitat/Range: Wet meadows, peat bogs, lake margins, and riverbanks in the lowland zone; rare in SW BC; E from AB to NF and S to NC, IL, and MT.

***Juncus breweri* Engelm.** (*J. lesuerii* [or *leseuerii*] auct. non Boland., *J. lesuerii* Bolander var. *tracyi* Jepson)
Brewer's rush
Habitat/Range: Mesic coastal sand dunes and open sandy beaches in the lowland zone; infrequent on Vancouver Island, rare on the Queen Charlotte Islands; S to CA.

***Juncus bufonius* L.**
Toad rush
Habitat/Range: Wet open places, disturbed soil, lakeshores, riverbanks and forest margins in all but the alpine zone; frequent throughout BC; cosmopolitan, N to AK, YT and NT, E to NF and S to FL, TX, AZ and CA; Eurasia; apparently both native and introduced in our area.
Notes: A variable species with a large number of chromosomal races. Several taxa within this complex have been described and treated as distinct species or varieties.

***Juncus bulbosus* L.** (*J. supinus* Moench)
Bulbous rush
Habitat/Range: In water in lakes and pools and wet peat bogs in the lowland zone; rare in SW BC; disjunct E to NF and NS and S to OR and N CA; Eurasia, N Africa; probably introduced in our area.

***Juncus castaneus* Sm.**
Chestnut rush
Habitat/Range: Wet marshes, peat bogs, river flats and margins of lakes and pools in the montane and subalpine zones; frequent in BC east of the Coast Mountains; circumpolar, N to AK, YT and NT, E to NF and S to CO and NM, Eurasia.
Notes: Two subspecies have been recognized in western North America. *J. castaneus* ssp. *leucochlamys* has not been reported from BC, but can be expected near the YT border.

1. Heads 1–3, dark, sessile; lowest bract shortest .... ssp. *castaneus*
1. Heads 3–5, pale, long peduncled; lowest bract longest .... ssp. *leucochlamys* (Zinz.) Hult.

***Juncus conglomeratus* L.** (*J. effusus* var. *conglomeratus* [L.] Engelm.)
Compact rush
Habitat/Range: Wet areas, disturbed soil and pastures in the lowland zone; rare in WC BC (reported by Hämet-Ahti [1980] from the Prince Rupert area); introduced from Europe.

***Juncus covillei* Piper**
Coville's rush
Habitat/Range: Wet lake margins and gravel banks in the lowland zone; rare on Vancouver Island and Calvert Island; S to CA.
Notes: The habit of this species is similar to *J. falcatus*. The most reliable identification character is the shape and size of seeds.

***Juncus drummondii* E. Meyer**
Drummond's rush
Habitat/Range: Dry meadows and rocky ridges in the subalpine and alpine zones; common throughout BC; N to AK, YT, and NT, E to AB and S to CO, NM and CA.
Notes: Two varieties occur in western NA. The var. *subtriflorus* can be expected in S BC.

1. Capsules usually at least 1 mm longer than the perianth; perianth 5–6 mm long .... var. *subtriflorus* (E. Meyer) C.L. Hitchc.
1. Capsules about equalling perianth; perianth 6–7 mm long .... var. *drummondii*

***Juncus effusus* L.**

Common rush

Habitat/Range: Mesic to wet disturbed places, pastures, fields, forest margins, marshes and peat bogs in the lowland, steppe and montane zones; common W of, and frequent E of the Coast-Cascade Mountains; cosmopolitan, N to AK, disjunct E from ON to PQ and NF, and S to FL, TX and CA; Eurasia, S Africa, S America, Australia.

Notes: A variable species with several varieties recognized in our area.

1. Perianth segments deep brown with green midvein, soft when dry; plant occurring throughout BC.
   2. Uppermost sheaths close, leathery, lustrous, and chestnut brown all over .... var. *gracilis* Hook.
   2. Uppermost sheaths looser, membranous, dull greenish-brown above .... var. *brunneus* Engelm.
1. Perianth segments greenish with little or no brownish border, stiff when dry; plants of coastal BC.
   3. Perianth 2–2.5 mm long; panicle compact, 1–4.5 cm long .... var *compactus* Lejeune & Court.
   3. Perianth 2.5–3.5 mm long; panicle loose, 2.5–15 cm long; .... var. *pacificus* Fern. & Wieg.

***Juncus ensifolius* Wikstr.** (*J. saximontanus* A. Nels. and *J. xiphoides* E. Meyer var. *montanus* Engelm. = var. *montanus*; *J. xiphoides* var. *triandrus* Engelm. = var. *ensifolius*)

Dagger-leaved rush

Habitat/Range: On wet open sandy soil, in marshes, and peat bogs, lakeshores and margins of creeks and rivers, from the lowland and steppe to subalpine zones; frequent in BC south of 55°N, infrequent northward; amphiberingian, N to AK, E to SK and disjunct in ON and PQ, and S to TX, CA and MX; E Asia.

Notes: A variable species. It includes two varieties sometimes treated as distinct species.

1. Heads 2–5, purplish brown; stamens 3; blades 3–6 mm broad; auricles absent .... var. *ensifolius*
1. Heads 5 or more, pale brown; stamens 6; blades 2–4 mm broad; auricles present .... var. *montanus* (Engelm.) C.L. Hitchc.

***Juncus falcatus* E. Meyer** (*J. falcatus* var. *sitchensis* Buch.)

Sickle-leaved rush

Habitat/Range: Wet rocky lake shores, streambanks and open soil in peat bogs in the lowland and montane zones; frequent in BC west of the Coast-Cascade Mountains; amphiberingian, N to AK and S to CA; E Asia and Australia.

***Juncus filiformis* L.**

Thread rush

Habitat/Range: Moist meadows and wet sandy lakeshores from the lowland and steppe to subalpine zones; infrequent throughout BC; circumpolar, N to AK, YT, and NT, E to NF and S to PA, WV, MI, MN, WY, UT, and OR; Eurasia, Patagonia.

***Juncus gerardii* Loisel. in Desv.**
Mud or Gerard's rush
Habitat/Range: Wet tidal flats in the lowland zone and along streams in the montane zone; infrequent in SW BC, rare in BC east of the Coast-Cascade Mountains (Anahim Lake area); introduced from Eurasia.

***Juncus kelloggii* Engelm.**
Kellogg's rush
Habitat/Range: Seasonally wet depressions and vernal pools in the lowland zone; rare on Vancouver Island, known only from Uplands Park, Oak Bay; S to CA.

***Juncus longistylis* Torr.**
Long-styled rush
Habitat/Range: Wet marshes and calcareous fens in the steppe and montane zones; infrequent in BC east of the Coast-Cascade Mountains; E to NF and S to MI, NE, NM, AZ and CA.

***Juncus mertensianus* Bong.**
Mertens' rush
Habitat/Range: Wet, grassy stream margins, snowbed sites, open boggy forests and disturbed areas from the montane to alpine zones; common throughout BC; amphiberingian, N to AK, YT and NT, E to AB and S to CO, NM and CA; Japan.

***Juncus nevadensis* S. Wats.** (*J. columbianus* Cov.)
Sierra rush
Habitat/Range: Wet lakeshores and stream margins in the lowland, steppe and montane zones; infrequent in the Cariboo-Chilcotin region, rare elsewhere; E to AB and S to NM, AZ, and CA.
Notes: This is a very variable species and several infraspecific taxa (often treated as distinct species) were described from WA and OR.

***Juncus nodosus* L.**
Tuberous rush
Habitat/Range: Wet lake shores and marshes in the lowland, steppe and montane zones; infrequent throughout BC, mostly in S BC east of the Coast-Cascade Mountains; N to AK, YT and NT, E to NF, and S to VA, IN, NE, NM and CA.

***Juncus oxymeris* Engelm.**
Pointed rush
Habitat/Range: Wet meadows and riverbanks in the lowland zone; rare in SW BC; S to AZ and CA.

***Juncus parryi* Engelm.**
Parry's rush
Habitat/Range: Dry meadows in the subalpine and alpine zones; frequent in BC south of 55°N, rare W of the Coast-Cascade Mountains; E to AB and S to UT, CO and CA.

***Juncus regelii* Buch.**
Regel's rush
Habitat/Range: Wet stream banks and marshes in the steppe and montane zones; rare E of the Coast-Cascade Mountains S of 55°N; E to AB and S to WY, NV and CA.

***Juncus stygius* L.** (*J. stygius* L. var. *americanus* Buch.)
Bog rush
Habitat/Range: Wet peat bogs in the lowland and montane zones; rare in coastal BC (N Vancouver Island and northern coast, N of Bella Bella); circumpolar, N to AK, YT and NT, E to NF and S to NY, MI and MN; Eurasia.
Notes: Plants from N America and E Asia have sometimes been separated as var. *americanus* Buch. The status of this variety is doubtful.

***Juncus supiniformis* Engelm.** (*J. oreganus* S. Wats.)
Spreading rush
Habitat/Range: Shallow water, wet muck, lakeshores and open bogs in the lowland and montane zones; common in BC west of the Coast-Cascade Mountains; N to AK and S to CA.

***Juncus tenuis* Willd.** (*J. confusus* Cov., *J. macer* S.F. Gray, *J. occidentalis* [Cov.] Wieg.)
Slender rush
Habitat/Range: Dry to moist pastures (mostly on disturbed soil) in the lowland, steppe and montane zones; common in BC south of 55°N, infrequent in N BC; N to AK, YT and NT, E to NF and S throughout US, C and S America.
Notes: A variable species. Several varieties have been recognized within this species; their taxonomy and distribution requires further study. Provisionally included in this complex are the closely related southern species *J. occidentalis* and *J. confusus*.

***Juncus torreyi* Cov.**
Torrey's rush
Habitat/Range: Moist river banks and lake shores in the steppe zone; infrequent in S BC east of the Coast-Cascade Mountains; E to PQ and S to MD, WV, MO, TX, NM and CA.

***Juncus triglumis* L.**
Three-flowered rush
Habitat/Range: Wet mossy areas and snowbed sites in the subalpine and alpine zones; infrequent in BC east of the Coast-Cascade Mountains and N of 51 1/2°N; circumpolar, N to AK, YT and NT, E to NF and S to UT and CO; Eurasia.

***Juncus vaseyi* Engelm.**
Vasey's rush
Habitat/Range: Dry to moist shores and meadows in the steppe and montane zones; infrequent in BC east of the Coast-Cascade Mountains; E to NF and S to ME, IL, CO and ID.

LUZULA

1. Inflorescence with solitary flowers or with flowers in groups of 2 or 3, not forming small heads.
    2. Flowers always single on long branches; leaves usually as long as stems or longer, pubescent on margins with long hairs; caruncle longer than 0.5 mm .... *L. rufescens*
    2. Flowers (at least most of them) usually in clusters of 2 or 3; leaves shorter than stems, glabrous or with hairs only around sheaths; caruncle no more than 0.3 mm long or absent.
        3. Perianth segments 2.5–3.0 mm long; anthers 1.2–1.5 mm long; capsules 2.8–3.8 mm long, ovate with beak up to 1 mm long .... *L. hitchcockii*
        3. Perianth segments usually shorter than 2.5 mm; anthers shorter than 1 mm long; capsules shorter than 2.5 mm, elliptical or rounded but never ovate, with a short or indistinct beak.
            4. Plants usually less than 30 cm tall; cauline leaves 3, up to 5 mm wide; bracts and bractlets strongly long ciliate .... *L. piperi*
            4. Plants usually more than 30 cm tall; cauline leaves (3) 4 or more, 5–10 mm wide; bracts entire to slightly ciliate, bractlets entire to lacerate but never strongly ciliate .... *L. parviflora*
1. Inflorescence with flowers in heads, spikes or in small clusters with more than 4 flowers.
    5. Bractlets scarcely fimbriate-ciliate, shorter than the flower; lower bract leaf-like, well developed.
        6. Inflorescence of 1 (3) sessile cluster of flowers .... *L. arctica*

6. Inflorescence of one subsessile and several ± pedunculate clusters of flowers.
    7. Inner perianth segments shorter than the outer ones; style much shorter than ovary .... *L. groenlandica*
    7. Inner and outer perianth segments of about equal length; style slightly shorter to longer than the ovary .... *L. multiflora*

5. Bractlets strongly fimbriate, nearly equalling and sometimes exceeding the flower; lower bract poorly developed.
    8. Inflorescence nodding, elongated spike, unbranched; the nerve of perianth segments continuous into an awn .... *L. spicata*
    8. Inflorescence head-like, or branched, upright; perianth segment not awned, at most shortly attenuate.
        9. Leaves ending with a callose thickening; basal sheaths light brown; plants densely cespitose .... *L. arctica*
        9. Leaves attenuate; basal sheaths purplish-brown; plants creeping or loosely cespitose.
            10. Outer cluster of flowers on recurved peduncles; perianth segments narrowly lanceolate; capsules ovate or elliptical, acute; sheaths at the base of leaves dull reddish; seeds 1–1.2 mm long, dark brown .... *L. arcuata*
            10. Outer clusters on erect peduncles or subsessile; perianth segments broadly lanceolate; capsules broadly obovate, mucronate; sheaths at the base of leaves shiny, reddish-brown; seeds up to 1.6 mm long, reddish brown .... *L. confusa*

***Luzula arctica* Blytt** (*L. nivalis* auct. non Laest. ex Sprengel)
Arctic woodrush
Habitat/Range: Moist meadows and snowbed sites in the subalpine and alpine zones; rare in SC and N BC; circumpolar, N to AK, YT, and NT and E to NF; Eurasia.

***Luzula arcuata* (Wahlenb.) Swallen** (*L. unalaschcensis* [Buch.] Satake, *L. arcuata* ssp. *unalaschcensis* [Buch.] Hult.)
Curved alpine woodrush
Habitat/Range: Dry to moist sites in the subalpine and alpine zones; infrequent in S BC, frequent in N BC, east of the Coast-Cascade Mountains; disjunct circumpolar, N to AK, YT and NT, E to AB and S to MT, ID and WA; Eurasia.

***Luzula confusa* Lindb.**
Confused woodrush
Habitat/Range: Moist rocky sites in the alpine zone; rare in NW BC; circumpolar, N to AK, YT and NT, E to NF, and S to ME and NH, Eurasia.

***Luzula groenlandica* Böcher**
Greenland woodrush
Habitat/Range: Wet depressions and snowbed sites in the alpine zone; rare in NW BC; N to AK and NT, E to NF and Greenland.

***Luzula hitchcockii* Hämet-Ahti** (*L. glabrata* auct. non [Hoppe] Desv.)
Small or Hitchcock's woodrush
Habitat/Range: Mesic to dry ridges and meadows in the subalpine and alpine zones; frequent in and E of the Coast-Cascade Mountains S of 52°N; E to AB and S to MT, ID or OR.

***Luzula multiflora* (Ehrh.) Lej.** (*L. campestris* [L.] DC. in DC. & Lam. var. *comosa* E. Meyer and var. *congesta* [Thuill.] E. Meyer = ssp. *comosa, L. campestris* var. *frigida* Buch. = ssp. *frigida, L. campestris* var. *multiflora* [Ehrh.] Célak. = ssp. *multiflora*)

Many-flowered woodrush

Habitat/Range: Dry open forests, rocky slopes and dry meadows in all zones; frequent throughout BC; circumpolar or disjunct cosmopolitan, N to AK, YT and NT, E to NF and S to VA, KY, IA, ND, MT, ID, and CA; Eurasia, S America, Philippines, New Zealand, Australia.

Notes: Three subspecies are known in BC, a fourth (ssp. *kjellmaniana*) may be expected.

1. Flowers in a congested head 1–3 cm long; flower clusters subsessile; dry open forests along the coast in SW BC .... ssp. *comosa* (E. Meyer) Hult.
1. Flowers in several clusters, at least some of the clusters stipitate.
   2. Outer perianth segments pointed, inner perianth segments truncate or obtuse mucronate; alpine zone in N BC .... ssp. *frigida* (Buch.) Krecz.
   2. All perianth segments equal in shape, slenderly pointed.
      3. Perianth segments indistinctly light-margined; yet to be found in BC, but expected in the alpine zone of NW BC .... ssp. *kjellmaniana* (Miyabe & Kudo) Tolm.
      3. Perianth segments prominently white-margined; grassy slopes in the lowland and montane zones, throughout BC; introduced from Europe .... ssp. *multiflora*

***Luzula parviflora* (Ehrh.) Desv.** (*L. divaricata* auct. non S. Wats. and *L. parviflora* ssp. *divaricata* [S. Wats.] Hult. = ssp. *fastigiata; L. parviflora* ssp. *melanocarpa* [Michx.] Hämet-Ahti nom. superfl. = ssp. *melanocarpa*)

Small-flowered woodrush

Habitat/Range: Mesic to moist alluvial forests in the lowland zone (ssp. *fastigiata*) and in the montane zone; common throughout BC; ssp. *melanocarpa* - N to AK and S to CA; ssp. *fastigiata* - circumpolar, N to AK, YT and NT, E to NF and S to NY, MI, WY and CA, Eurasia, Greenland.

Notes: Two subspecies can be distinguished in our area. *Luzula parviflora* ssp. *fastigiata* may be better treated as a distinct species.

1. Stem with distinctly reddish colour at the lowermost internode; stem leaves lanceolate, acuminate, green, shiny and thin; perianth segments very acute, scarcely translucent with distinct middle nerve; capsule pale brown to straw-coloured .... ssp. *fastigiata* (E. Meyer) Hämet-Ahti
1. Stems without reddish colour or with indistinct reddish tinge at very base; cauline leaves broadly linear to lanceolate, acute, bluish or greyish green, dull, thick to thin; perianth segments translucent, middle nerve fairly distinct; capsule usually blackish brown .... ssp. *melanocarpa* (Michx.) Tolm.

***Luzula piperi* (Cov.) M.E. Jones** (*L. wahlenbergii* auct. non Rupr., *L. wahlenbergii* Rupr. ssp. *piperi* [Cov.] Hult.)

Piper's woodrush

Habitat/Range: Mesic to moist meadows, snowbed sites and shrublands in the subalpine and alpine zones; frequent throughout BC in and E of the Coast-Cascade Mountains; amphiberingian, N to AK and YT, E to AB and S to CO and N CA, E Asia.

***Luzula rufescens* Fischer ex E. Mey.**

Rusty woodrush

Habitat/Range: Dry to mesic open forests and forest margins in the montane zone; rare in NE BC; amphiberingian, N to AK and YT; E Asia.

***Luzula spicata* (L.) DC.**
Spiked woodrush
Habitat/Range: Dry rocky slopes in the subalpine and alpine zones; frequent throughout BC; circumpolar, N to AK, YT and NT, disjunct E to MB to NF, and S to NY and to CO, AZ and CA; Eurasia, Greenland.

JUNCAGINACEAE

TRIGLOCHIN

1. Carpels and stigmas 3; fruits linear or slenderly club-shaped .... *T. palustre*
1. Carpels and stigmas 6; fruits oblong or ovoid-prismatic.
    2. Ligules entire or nearly so, (1) 1.5–5 mm long; plants (1) 3–10 (12) dm tall .... *T. maritimum*
    2. Ligules bilobed, 0.5–1 mm long; plants 0.5–1.5 (2) dm tall .... *T. concinnum*

Triglochin concinnum* Davy var. *concinnum
Graceful arrow-grass
Habitat/Range: Tidal marshes; rare in coastal BC; S to NW MX; S America.

***Triglochin maritimum* L.**
Seaside arrow-grass
Habitat/Range: Tidal marshes, saline and alkaline ponds and wet sites in the lowland, steppe and montane zones; common throughout BC south of 55°N, less frequent northward; cosmopolitan, N to AK, YT and NT, E to NF and S to TX, NM and MX; Eurasia, N Africa, S America.

***Triglochin palustre* L.**
Marsh arrow-grass
Habitat/Range: Tidal marshes and saline and alkaline ponds and wet sites in the lowland, steppe and montane zones; common in BC east of the Coast-Cascade Mountains and S of 55°N, less frequent northward, rare along the coast; circumpolar, N to AK, YT and NT, E to NF and S to OH, PA, NE, NM and CA; Greenland, Eurasia, S America.

LEMNACEAE

1. Fronds unnerved, 0.5–1.5 mm long, rootless .... *Wolffia*
1. Fronds nerved, 2.0 mm or more long, with roots.
    2. Fronds usually 3-nerved, with solitary roots .... *Lemna*
    2. Fronds 4–12-nerved, with 5–15 roots .... *Spirodela*

LEMNA

1. Fronds oblong to lanceolate, 6–12 mm long, long-stalked, remaining attached in a zig-zag chain, usually submersed .... *L. trisulca*
1. Fronds elliptic-obovate to round-ovate, 2–6 mm long, sessile, quickly becoming detached, floating on the surface .... *L. minor*

***Lemna minor* L.** (*L. turionifera* Landolt)

Common duckweed or water lentil

Habitat/Range: Ponds and slow-moving streams in the lowland, steppe and montane zones; common throughout BC south of 55°N, less frequent northward, absent in NW BC and the Queen Charlotte Islands; circumpolar, N to AK, YT and NT, E to NF and S to FL, TX, CA and MX; Eurasia.

Notes: The separation of *L. turionifera* from *L. minor*, based on characters which are often absent or are sometimes found in *L. minor*, does not appear justified.

***Lemna trisulca* L.**

Ivy-leaved or star duckweed

Habitat/Range: Ponds and slow-moving streams in the lowland, steppe and montane zones; frequent throughout BC, absent on the Queen Charlotte Islands; circumpolar, N to AK, YT and NT, E to NB and NS and S to FL, TX, CA and MX; Eurasia, S America.

SPIRODELA

***Spirodela polyrhiza* (L.) Schleid.**

Great duckweed

Habitat/Range: Ponds and slow-moving streams in the lowland, steppe and montane zones; common in SW and SC BC, infrequent elsewhere except absent on N Vancouver Island, the Queen Charlotte Islands and NW BC; circumpolar, E to PE and NS and S to FL, TX, CA and MX; Eurasia, S America.

WOLFFIA

1. Plants ellipsoid, nearly twice as long as wide .... *W. borealis*
1. Plants globose, about as long as wide .... *W. columbiana*

***Wolffia borealis* (Engelm. ex Hegelmaier) Landolt & Wildi** (*W. punctata* Griseb. pro parte)

Water-meal

Habitat/Range: Ponds, lakes and slow-moving streams in the lowland and montane zones; rare in the lower Fraser Valley and SE BC (Creston); disjunct in ON and S to NY, MI, IA and OH.

***Wolffia columbiana* Karst.**

Water-meal

Habitat/Range: Ponds and lakes in the lowland zone; rare in SW BC, known only from Beaver, Blenkinsop and Swan Lakes; disjunct E to ON and PQ and S to much of US; S. America

LILAEACEAE[10]

LILAEA

***Lilaea scilloides* (Poir.) Haum.** (*L. subulata* Humboldt & Bonpland)

Flowering quillwort

Habitat/Range: Mud flats, ponds and wet sites in the lowland zone; rare in coastal BC; E to S SK and S to MT, NV and MX; S America.

---

10 Sometimes included in Juncaginaceae.

LILIACEAE[11]

1. Flowers 2-merous .... *Maianthemum*
1. Flowers 3-merous.
    2. Leaves scale-like, subtending soft, needle-like, green branches (phyllodes) .... *Asparagus*
    2. Leaves not scale-like, broader, phyllodes absent.
        3. Ovaries distinctly inferior .... *Narcissus*
        3. Ovaries superior (rarely partially inferior).
            4. Leaves in a single whorl of three .... *Trillium*
            4. Leaves not in a single whorl of three.
                5. Perianth segments in two very different series; the petals much broader than the sepals .... *Calochortus*
                5. Perianth segments more or less alike.
                    6. Flowers borne in umbels.
                        7. Plants with the odour of onions; perianth segments distinct to base .... *Allium*
                        7. Plants lacking odour of onions; perianth segments connate up to half their length.
                            8. Anthers 6, attached in middle .... *Triteleia*
                            8. Anthers 3, attached at base .... *Brodiaea*
                    6. Flowers solitary, racemose or paniculate, not umbellate.
                        9. Leaves tough, persisting for more than one year, as large basal tuft .... *Xerophyllum*
                        9. Leaves soft, not persisting for more than one year.
                            10. Leaves mostly on the flowering stems, not strongly reduced upward.
                                11. Plants from scaly bulbs; leaves often whorled.
                                    12. Tepals 4–10 cm long, bright yellow to orange .... *Lilium*
                                    12. Tepals 1–3.5 cm long, dark green to brown or black .... *Fritillaria*
                                11. Plants from rhizomes; leaves alternate, never whorled.
                                    13. Flowers more than 10 in terminal racemes or panicles.
                                        14. Styles 3; tepals 6–17 mm long .... *Veratrum*
                                        14. Styles 1; tepals 2–7 mm long .... *Smilacina*
                                    13. Flowers borne singly along stems or in clusters of 1–3 at tips.
                                        15. Flowers 1–2 along lower side of stems .... *Streptopus*
                                        15. Flowers 1–3 at tips of main stems .... *Disporum*

---

[11] Key modified after Hitchcock and Cronquist (1973).

10. Leaves primarily basal; flower stems naked, or with leaves greatly reduced upward.

16. Leaves with broad white mid-vein; tepals with green stripe on outside .... *Ornithogalum*

16. Leaves plain green, without white mid-vein; tepals lacking green stripe on outside.

17. Styles 3, distinct almost to the ovary.

18. Leaves 2-ranked and equitant; plants rhizomatous .... *Tofieldia*

18. Leaves neither 2-ranked nor equitant; plants bulbous.

19. Tepals with a prominent gland near their bases; flowers rotate to shallowly campanulate .... *Zygadenus*

19. Tepals lacking glands; flowers tubular-campanulate.

20. Tepals greenish-yellow to purplish-green; plants montane to alpine ...... *Stenanthium*

20. Tepals blue, pink or white; plants of the lowland zone .... *Hyacinthoides*

17. Styles 1, or 3, but connate more than half their length.

21. Leaves several, linear to linear-lanceolate, at least 10 times as long as broad.

22. Flowers borne singly or at most in 2's; plants not more than 15 cm tall, subalpine to alpine .... *Lloydia*

22. Flowers in racemes; plants more than 20 cm tall, of the lowlands .... *Camassia*

21. Leaves usually 2, lanceolate or oblong to oblanceolate, less than 10 times as long as broad.

23. Plants rhizomatous; filaments pubescent; fruits fleshy blue berries at maturity .... *Clintonia*

23. Plants cormous; filaments glabrous; fruits dry capsules at maturity .... *Erythronium*

ALLIUM

1. Inner filaments with long appendages, longer than the anthers; plants introduced weeds of SW BC ... . *A. vineale*

1. Inner filaments lacking appendages; plants native.

2. Inflorescences more or less capitate, with very short pedicels, flowering from the centre outward; leaves terete .... *A. schoenoprasum*

2. Inflorescences umbellate, flowering from the outside inward; leaves more or less flattened.

3. Bulb coats persisting as a network of reticulate fibers .... *A. geyeri*

3. Bulb coats not of reticulate fibres.
    4. Plants from distinctive horizontal rhizomes .... *A. validum*
    4. Plants from bulbs, lacking distinctive rhizomes.
        5. Bulbs elongated; umbels usually nodding; stamens exserted .... *A. cernuum*
        5. Bulbs ovoid; umbels usually erect; stamens included.
            6. Leaves usually two, curved to one side, longer than the scapes .... *A. crenulatum*
            6. Leaves usually more than two, not curved to one side, shorter than the scapes.
                7. Perianth segments long-acuminate, with recurved tips ...... *A. acuminatum*
                7. Perianth segments elliptic-lanceolate, tips not recurved ...... *A. amplectens*

Allium acuminatum **Hook.**
Hooker's onion
Habitat/Range: Dry rocky bluffs and meadows in the lowland and steppe zones; infrequent on SE Vancouver Island, the Gulf Islands and the Fraser Valley, north to Lytton and Pemberton; S to CA, AZ, WY and CO.

Allium amplectens **Torr.**
Slimleaf onion
Habitat/Range: Dry rocky bluffs and meadows in the lowland zone; rare on SE Vancouver Island, the Gulf Islands and the adjacent mainland (Powell River); S to CA.

Allium cernuum **Roth in Roem.**
Nodding onion
Habitat/Range: Mesic to moist rocky bluffs, open forests and meadows in the lowland and steppe to subalpine zones; common in BC south of 56°N; E to ON and S to NY, MI, AZ and OR.

Allium crenulatum **Wieg.**
Olympic onion
Habitat/Range: Dry rock outcrops and rockslides in the lowland and alpine zones; rare on Vancouver Island and the Gulf Islands; S to SW OR.

Allium geyeri **S. Wats.**
Geyer's onion
Habitat/Range: Moist meadows and rock outcrops in the lowland, steppe and montane zones; rare on Vancouver Island and the Fraser and Thompson River valleys; E to SK and S to WY, NV and NM.
Notes: Two varieties occur in BC.

1. Heads with all flowers functional .... var. *geyeri*
1. Heads with at least some flowers replaced by bulbils .... var. *tenerum* M.E. Jones

Allium schenoprasum **L. var.** ***sibiricum*** **(L.) Hartm.**
Wild chives
Habitat/Range: Wet meadows, streambanks and lake margins in the montane zone; infrequent in N and E BC; circumboreal, N to AK, YT and NT, E to NF and S to NY, MN, CO and OR; Eurasia.

Allium validum **S. Wats.**
Swamp onion
Habitat/Range: Moist meadows in the steppe to alpine zones; rare in SC and SE BC (known only from Okanagan Falls, Corbin and Kootenay National Park); S to ID, NV and CA.

Allium vineale **L.**

Field garlic

Habitat/Range: Fields and waste places in the lowland zone; infrequent on S Vancouver Island; introduced from Europe.

ASPARAGUS

Asparagus officinalis **L.**

Garden asparagus

Habitat/Range: Dry to mesic fields, floodplain thickets, roadsides and ditches in the lowland and steppe zones; common in S BC east of the Coast-Cascade Mountains, rare along the SW coast; introduced from Europe.

BRODIAEA

Brodiaea coronaria **(Salisb.) Engl.**

Harvest brodiaea

Habitat/Range: Dry to mesic grassy slopes and rocky bluffs in the lowland and steppe zones; infrequent on SE Vancouver Island, the Gulf Islands and adjacent mainland (isolated population at Spences Bridge); S to CA.

CALOCHORTUS

1. Petals pale pink to purple, 3.0–4.5 cm long; sepals longer than petals .... *C. macrocarpus*
1. Petals white or yellowish-white, less than 2.5 cm long; sepals shorter than petals.
   2. Capsules nodding in fruit; petals with circular gland; anthers long apiculate .... *C. apiculatus*
   2. Capsules erect in fruit; glands crescent-shaped; anthers acute or acuminate, not long apiculate .. .. *C. lyallii*

Calochortus apiculatus **Baker**

Three-spot mariposa lily

Habitat/Range: Dry to mesic meadows and forests in the steppe and montane zones; infrequent in SE BC; E to AB and S to MT, ID and WA.

Calochortus lyallii **Baker**

Lyall's mariposa lily

Habitat/Range: Dry to mesic forests and slopes in the steppe and montane zones; rare in SC and S to WA.

Calochortus macrocarpus **Dougl.**

Sagebrush mariposa lily

Habitat/Range: Dry sites in shrublands and grasslands (Fraser Plateau) in the steppe zone; frequent in SC BC; S to MT, NV and CA.

CAMASSIA

1. Perianth segments twisted together around ovaries after anthesis; flowers regular; fruiting pedicels longer than the bracts .... *C. leichtlinii*
1. Perianth segments not twisted together around ovaries after anthesis; flowers irregular, with 5 perianth segments curving upward and 1 pointing downward; fruiting pedicels shorter than the bracts .... *C. quamash*

Camassia leichtlinii **(Baker) S. Wats.**
Great camas
Habitat/Range: Vernally moist meadows and open woodlands in the lowland zone; locally frequent on S Vancouver Island and the Gulf Islands; S to CA.

Camassia quamash **(Pursh) Greene**
Common camas
Habitat/Range: Dry to mesic or vernally moist meadows in the lowland, steppe and montane zones; locally frequent on SE Vancouver Island, the Gulf Islands and SE BC; E to AB and S to MT, WY and UT.
Notes: Two varieties occur in BC.

1. Perianth segments 3–5-veined, mostly less than 6 mm wide, usually less than 30 mm long; pedicels appressed in fruit; plants from the dry interior .... var. *quamash*
1. Perianth segments 5–9-veined, mostly more than 6 mm wide, usually more than 30 mm long; pedicels ascending or spreading in fruit; plants from the coast .... var. *maxima* (Gould) Boivin

CLINTONIA

Clintonia uniflora **(Schult.) Kunth**
Queen's cup, or blue-bead clintonia
Habitat/Range: Moist to mesic forests in the lowland to lower subalpine zones; common in BC south of 56°N; N to S AK, E to SW AB and S to MT, ID and CA.

DISPORUM

1. Flowers 15–28 mm long; stamens concealed by tepals; tepals flared only at tips .... *D. smithii*
1. Flowers 8–10 mm long; stamens exserted, or at least not concealed by tepals; tepals broadly flared.
   2. Leaves usually hairy on upper surface; cilia on leaf margins pointed forward; ovaries ellipsoid, pointed; styles usually hairy; fruits smooth or hairy, but not papillose, 4–6-seeded ...... *D. hookeri*
   2. Leaves glabrous on upper surface; cilia on leaf margins spreading; ovaries rounded; styles glabrous; fruits papillose, 6–15-seeded .... *D. trachycarpum*

Disporum hookeri **(Torr.) Nichols. var. *oreganum* (S. Wats.) Q. Jones**
Hooker's fairybells
Habitat/Range: Moist to mesic forests in the lowland to lower subalpine zones; common in BC south of 55°N; E to AB and S to ID, W MT and NW OR.

Disporum smithii **(Hook.) Piper**
Smith's fairybells
Habitat/Range: Moist forests and streambanks in the lowland zone; rare on SW Vancouver Island; S to CA.

Disporum trachycarpum (S. Wats.) Benth. & Hook.
Rough-fruited fairybells
Habitat/Range: Moist to mesic wooded slopes in the steppe and montane zones; common in BC east of the Coast-Cascade Mountains and S of 56°N; E to AB and S to ND, CO, AZ and OR.

ERYTHRONIUM

1. Leaves mottled with green and dark brown; flowers white or pink; plants of low elevations.
   2. Flowers white ..... *E. oregonum*
   2. Flowers pink ..... *E. revolutum*
1. Leaves bright yellow-green, not mottled; flowers white or bright yellow; plants of high elevations.
   3. Flowers bright yellow ..... *E. grandiflorum*
   3. Flowers white ..... *E. montanum*

Erythronium grandiflorum Pursh
Yellow glacier lily or avalanche lily
Habitat/Range: Moist meadows, edges of snow-melt and along streams in the montane to alpine zones; common in BC S of 53°N; E to AB and S to MT, WY, UT and OR.

Erythronium montanum S. Wats.
White glacier lily
Habitat/Range: Moist meadows and open forests in the montane to alpine zones; rare on Vancouver Island and the coastal mainland; S to N OR.

Erythronium oregonum Appleg.
White fawn lily
Habitat/Range: Dry to mesic meadows and forests in the lowland zone; locally common on SE Vancouver Island, the Gulf Islands and the adjacent mainland; S to S OR.

Erythronium revolutum J.E. Sm. in Rees
Pink fawn lily
Habitat/Range: Moist forests and riverbanks in the lowland zone; locally frequent on Vancouver Island and the adjacent mainland; S to N CA.

FRITILLARIA

1. Flowers bright yellow, fading to orange; stigmas entire; stamens about as long as tepals ... *F. pudica*
1. Flowers dull green to dark purplish-brown; stigmas deeply lobed; stamens about half as long as tepals.
   2. Flowers not mottled; tepals ridged internally along veins; capsules not winged ..... *F. camschatcensis*
   2. Flowers mottled or spotted; tepals not ridged internally; capsules winged ..... *F. affinis*

Fritillaria affinis (Schultes) Sealy (*F. lanceolata* Pursh)
Chocolate lily
Habitat/Range: Dry to mesic grassy bluffs, meadows and open woodlands in the lowland and steppe to lower subalpine zones; common on SE Vancouver Island, the Gulf Islands and inland to SC BC, south of 52°N; S to ID and CA.
Notes: This species has long been known as *F. lanceolata* Pursh, an illegitimate name.

Fritillaria camschatcensis **(L.) Ker-Gawl.**
Northern rice-root, or riceroot fritillary
Habitat/Range: Moist tideflats, meadows, open forests and streambanks in the lowland subalpine zones; common along coast, infrequent inland; N to AK and S to W WA.

Fritillaria pudica **(Pursh) Spreng.**
Yellow bell, or yellow fritillary
Habitat/Range: Dry shrublands, grasslands and open coniferous forests in the steppe zone; infrequent in S BC east of the Coast-Cascade Mountains; E to AB and S to WY, UT, NV and N CA.

HYACINTHOIDES

Hyacinthoides hispanica **(Mill.) Roth.** (*Endymion hispanicus* [Mill.] Chouard, *Scilla hispanica* Mill.)
Spanish bluebells
Habitat/Range: Edges of forests, roadsides and waste places in the lowland zone; locally frequent garden escape on S Vancouver Island, the Gulf Islands and the adjacent mainland (one record from the Queen Charlotte Islands); introduced from Europe.

LILIUM

1. Flowers usually several in a raceme, nodding; tepals yellow-orange, strongly recurved .... *L. columbianum*
1. Flowers 1–3, solitary to sub-umbellate, erect; tepals orange-red to brick-red, not recurved .... *L. philadelphicum*

Lilium columbianum **Hanson ex Baker** (*L. canadense* L. var. *parviflorum* Hook.)
Tiger or columbia lily
Habitat/Range: Moist open forests, roadsides and meadows in the steppe to alpine zones; common throughout S BC south of 54°N; S to ID, NV and N CA.

Lilium philadelphicum **L. var.** ***andinum*** **(Nutt.) Ker-Gawl.**
Wood lily
Habitat/Range: Mesic meadows and edges of forests (often on limestone) in the steppe and montane zones; infrequent in the Peace River region and SE BC; E to SK and S to ME, WV, NV and NM.

LLOYDIA

Lloydia serotina **(L.) Richenb.** (*L. serontina* ssp. *flava* Calder & Taylor = var. *flava*)
Alp lily
Habitat/Range: Mesic rocky slopes, rock crevices and gravelly meadows in the alpine zone; infrequent in the Queen Charlotte Islands (var. *flava*) and frequent in scattered localities throughout W BC (var. *serotina*); circumboreal, N to AK, YT and NT, E to NF and S to NV, NM and NW OR.
Notes: Two varieties occur in BC.

1. Tepals with purple veins; basal leaves often shorter than the flowering stems ... var. *serotina*
1. Tepals with yellow or green veins; basal leaves often longer than flowering stems .... var. *flava* (Calder & Taylor) Boivin

MAIANTHEMUM

1. Leaves sub-sessile, the blades with ciliate margins and pubescent beneath .... *M. canadense*
1. Leaves distinctly petiolate, the blades glabrous .... *M. dilatatum*

***Maianthemum canadense* Desf. var. *interius* Fern.**
Canadian mayflower, or wild lily-of-the-valley
Habitat/Range: Moist to mesic forests in the montane zone; infrequent in BC east of the Coast Mountains, north of 52°N; N to NT, E to NF and S to GA, TN, IA and ND.

***Maianthemum dilatatum* (A. Wood) Nels. & Macbr.**
False lily-of-the-valley, or two-leaved false Solomon's-seal
Habitat/Range: Moist to mesic forests and streambanks in the lowland and montane zones; common in coastal BC; N to AK and S to ID and CA.

NARCISSUS

1. Tepals predominately yellow; corona at least half as long as tepals .... *N. pseudonarcissus*
1. Tepals white; corona much less than half as long as tepals .... *N. poeticus*

***Narcissus poeticus* L.**
Poet's narcissus
Habitat/Range: Fields, roadsides and waste places in the lowland zone; rare garden escape on S Vancouver Island; introduced from Europe.

***Narcissus pseudonarcissus* L.**
Daffodil
Habitat/Range: Fields, edges of forests, roadsides and waste places in the lowland zone; rare garden escape on S Vancouver Island, the Gulf Islands and the lower mainland; introduced from Europe.
Notes: Probably not truly naturalized, but long persisting and multiplying vegetatively by bulblets. Most of the plants found apparently naturalized in BC are probably hybrids or cultivated forms of this species and not the wild species itself.

ORNITHOGALUM

***Ornithogalum umbellatum* L.**
Star-of-Bethlehem
Habitat/Range: Roadsides, fields and waste places in the lowland zone; rare garden escape, but apparently re-seeding and established on S Vancouver Island and the adjacent mainland; introduced from Europe.

SMILACINA

1. Flowers many, subsessile in large panicles .... *S. racemosa*
1. Flowers few, distinctly pedicellate, in racemes.
    2. Leaves more than 4, sessile, abruptly narrowed at their base .... *S. stellata*
    2. Leaves 2–4, tapering to somewhat sheathing bases .... *S. trifolia*

Smilacina racemosa **(L.) Desf. var.** ***amplexicaulis*** **(Nutt. ex Baker) S. Wats.**

False Solomon's-seal

Habitat/Range: Moist to dry forests, rocky places and streambanks in the lowland, steppe and montane zones; common throughout BC south of 56°N; N to AK, E to NS and PQ and S to GA, MO, AZ and CA.

Smilacina stellata **(L.) Desf.**

Star-flowered false Solomon's-seal

Habitat/Range: Moist to mesic forests, streambanks and meadows in the montane to alpine zones; common throughout BC; N to AK, YT and NT, E to NF and S to VA, OH, KS, NM and CA.

Smilacina trifolia **(L.) Desf.**

Three-leaved false Solomon's-seal

Habitat/Range: Moist forests, bogs and peaty streambanks in the montane and subalpine zones; infrequent in N BC, north of 56°N; N to AK, YT and NT, E to NF and S to PA, IL and MN.

STENANTHIUM

Stenanthium occidentale **A. Gray**

Western mountainbells

Habitat/Range: Damp forests and boggy meadows in the montane to alpine zones; frequent in SW and SE BC north to 52°N; E to AB and S to MT, ID and CA.

STREPTOPUS

1. Plants usually branched; leaves with clasping bases; leaf margins not ciliate; stems not pubescent at the nodes; tepals greenish-white .... *S. amplexifolius*
1. Plants usually unbranched; leaves not clasping; leaf margins ciliate; stems pubescent at the nodes; tepals rose.
    2. Flowers shallowly bowl-shaped, the lobes wide-spreading; tepals about 2.5 mm long .... *S. streptopoides*
    2. Flowers campanulate, the lobes recurved only at the tips; tepals about 10 mm long .... *S. roseus*

Streptopus amplexifolius **(L.) DC. in Lam. & DC.** (*S. amplexifolius* var. *denticulatus* Fassett = var. *americanus*)

Clasping twistedstalk

Habitat/Range: Moist forests in the lowland and steppe to subalpine zones; common throughout BC; circumboreal, N to AK, E to NF and S to NC, NM and CA; Greenland, Eurasia.

Notes: Two varieties occur in BC.

1. Stems coarsely pubescent below the first branches; leaves with minute denticulations; plants mostly from west of the Coast-Cascade Mountains .... var. *americanus* Schult.
1. Stems glabrous; leaf margins entire; plants mostly from east of the Coast-Cascade Mountains .... var. *chalazatus* Fassett

Streptopus roseus **Michx. var.** ***curvipes*** **(Vail) Fassett**

Rosy twistedstalk

Habitat/Range: Moist forests in the lowland and steppe to subalpine zones; frequent throughout BC south of 56°N; N to AK.

Streptopus streptopoides **(Ledeb.) Frye & Rigg. var.** ***brevipes*** **(Baker) Fassett**
Small twistedstalk
Habitat/Range: Moist forests in the montane and subalpine zones; common on coastal mainland, infrequent elsewhere in BC south of 56°N; amphiberingian, N to AK and S to WA; E Asia.

TOFIELDIA

1. Stems glandular pubescent, at least in the inflorescences .... *T. glutinosa*
1. Stems glabrous throughout.
    2. Stems leafless or at most with one small bract; flowers greenish-white .... *T. pusilla*
    2. Stems with one or more leaves; flowers purplish .... *T. coccinea*

Tofieldia coccinea **Richards.**
Northern false asphodel
Habitat/Range: Dry to mesic meadows and rocky slopes in the montane to alpine zones; rare in N BC; circumpolar, N to AK, YT and NT and E to NF; Greenland, Eurasia.

Tofieldia glutinosa **(Michx.) Pers.**
Sticky false asphodel
Habitat/Range: Wet meadows and bogs in the montane to alpine zones; common throughout BC; N to AK, YT and NT, E to NF and S to NC, OH, IN, MN and CA.
Notes: A number of infraspecific taxa, not treated here, have been recognized within this species. For a discussion of these see Hitchcock (1944).

Tofieldia pusilla **(Michx.) Pers.** (*T. borealis* Wahlenb.)
Common false asphodel
Habitat/Range: Moist, gravelly open areas and marshy meadows in the montane to alpine zones; infrequent in E and N BC; circumpolar, N to AK, YT and NT, E to NF and S to MT; Greenland, Eurasia.

TRILLIUM

Trillium ovatum **Pursh**
Western or white trillium
Habitat/Range: Moist forests in the lowland, steppe and montane zones; common in S BC, south of 50°N; E to SW AB and S to CO and CA.
Notes: Small plants found rarely on western Vancouver Island, with flowers opening pink, have been named forma *hibbersonii* Taylor & Szczawinski. Considered by some to be a distinct species, they need further study, especially in relation to other small forms found elsewhere in the range of the species.

TRITELEIA

1. Flowers white, shallowly campanulate or rotate; tepals not concealing the ovaries .... *T. hyacinthina*
1. Flowers pale to dark blue, turbinate; tepals concealing the ovaries.
    2. Filaments flat, broader than anthers, unequal in length, but inserted at same level ...... *T. howellii*
    2. Filaments not flat, narrower than anthers, unequally inserted .... *T. grandiflora*

Triteleia grandiflora **Lindl.** (*Brodiaea douglasii* [Lindl.] S. Wats.)
Large-flowered triteleia
Habitat/Range: Dry slopes in the steppe zone; frequent in SC and SE BC; S to WY, UT and OR.
Notes: Formerly known from S Vancouver Island, but probably extirpated there.

Triteleia howellii **(S. Wats.) Greene** (*Brodiaea howellii* S. Wats.)
Howell's triteleia
Habitat/Range: Dry to mesic grassy coastal bluffs in the lowland zone; rare on S Vancouver Island, the Gulf Islands, and the adjacent mainland; S to OR.

Triteleia hyacinthina **(Lindl.) Greene** (*Brodiaea hyacinthina* [Lindl.] Baker)
White triteleia, or fool's-onion
Habitat/Range: Dry to mesic grassy coastal bluffs, rocky slopes and meadows in the lowland zone; frequent on S Vancouver Island, the Gulf Islands and the adjacent mainland; S to CA.

VERATRUM

Veratrum viride **Ait. ssp.** ***eschscholtzii*** **(A. Gray) Löve & Löve** (*V. eschscholtzii* A. Gray)
Indian or green false hellebore
Habitat/Range: Moist meadows, streambanks, swamps and open forests in the lowland to subalpine zones; common throughout BC, especially southward; N to AK, E to PQ and S to NC, ID, MT and OR.

XEROPHYLLUM

Xerophyllum tenax **(Pursh) Nutt.**
Bear-grass
Habitat/Range: Dry coniferous forests and clearings in the montane zone; locally common in extreme SE BC; S to ID, MT and CA.

ZYGADENUS

1. Perianth segments 8–10 mm long; ovaries half inferior; floral glands distinctly bilobed .... *Z. elegans*
1. Perianth segments less than 7 mm long; ovaries superior; floral glands less distinctive and unlobed .... *Z. venenosus*

Zygadenus elegans **Pursh**
Mountain or elegant death-camas
Habitat/Range: Damp meadows and open forests (often on calcareous sites) in the steppe to alpine zones; frequent throughout BC east of the Coast-Cascade Mountains, more common in N BC; N to AK, E to AB and S to SD, CO, TX, N MX and OR.

Zygadenus venenosus **S. Wats.** (*Z. gramineus* Rydb. = var. *gramineus*)
Meadow or grass-leaved death-camas
Habitat/Range: Dry to mesic grassy hillsides and meadows in the lowland to alpine zones; common in S BC (var. *venenosus* predominantly along coast and var. *gramineus* more common in the interior), S of 51°N; E to SW SK and S to NE, CO and Baja CA.
Notes: Two varieties occur in BC.

1. Upper stem leaves with papery sheaths; 3 outer tepals sessile or short-clawed, only the inner 3 long-clawed .... var. *gramineus* (Rydb.) Walsh
1. Upper stem leaves sheathless, all 6 tepals equally clawed .... var. *venenosus*

NAJADACEAE

NAJAS

***Najas flexilis* (Willd.) Rostk. & Schmidt**
Wavy water nymph
Habitat/Range: Ponds and marshes in the lowland, steppe and montane zones; common in S BC; circumpolar, N to NT, E to NF and S to NC, IA, SD and CA; Eurasia.

ORCHIDACEAE

1. Plants saprophytic, lacking green stems and leaves.
   2. Stems yellow to red-brown; underground parts coral-like; lips not sac-like .... *Corallorhiza*
   2. Stems white; underground parts not coral-like; lips sac-like at base .... *Cephalanthera*
1. Plants not saprophytic, bearing green stems and leaves (leaves may be withered by anthesis).
   3. Lips a large inflated sac more than 1 cm long.
      4. Leaves two or more (either basal or on the flowering stems) .... *Cypripedium*
      4. Leaves one (basal) .... *Calypso*
   3. Lips not sac-like, or if so, less than 1 cm long.
      5. Flowers with a distinct spur projecting downward from the base of lip.
         6. Flowers white to pale pink, spotted with dark red .... *Amerorchis*
         6. Flowers white to green, not spotted.
            7. Lips three-lobed at tips; bracts greatly exceeding flowers .... *Coeloglossum*
            7. Lips not lobed at tips; bracts not usually exceeding flowers.
               8. Sepals 1-nerved; basal leaves usually withered at flowering time .... *Piperia*
               8. Sepals 3-nerved; basal leaves remaining green at flowering time .... *Platanthera*
      5. Flowers lacking a spur.
         9. Leaves basal.
            10. Leaves fleshy and evergreen, mottled or veined with white; plants not cormous .... *Goodyera*
            10. Leaves membraneous, not evergreen and not mottled or veined with white; plants cormous.
                11. Flowers and mature capsules about 1 cm long .... *Liparis*
                11. Flowers and mature capsules less than 0.5 cm long.
                    12. Leaves one (rarely two), 3-6 cm long, prominently several nerved .... *Malaxis*
                    12. Leaves two or three, less than 2 cm long, indistinctly few nerved .... *Hammarbya*
         9. Leaves, at least some of them, cauline.
            13. Leaves two, opposite and sessile .... *Listera*

13. Leaves several, alternate.
    14. Leaves not much reduced upward; bracts subtending flowers leafy; flowers brown, large, more than 1 cm long .... *Epipactis*
    14. Leaves strongly reduced upward; bracts not leaf-like; flowers white to green, small, less than 1 cm long .... *Spiranthes*

AMERORCHIS

***Amerorchis rotundifolia* (Banks ex Pursh) Hult.** (*Orchis rotundifolia* Banks ex Pursh)
Round-leaved orchis
Habitat/Range: Moist, mossy forests, swamps and bogs in the lowland zone; infrequent throughout BC east of the Coast-Cascade Mountains; N to AK, YT and NT, E to NF and S to VT, WI, WY and MT; Greenland.

CALYPSO

***Calypso bulbosa* (L.) Oakes in Thomps.** (*C. bulbosa* ssp. *occidentalis* [Holzinger] Calder & Taylor)
Fairyslipper
Habitat/Range: Dry to moist mossy forests in the lowland, steppe and montane zones; common throughout BC; circumpolar, N to AK, E to NF and S to MA, MI, CO, AZ and CA; Eurasia.

CEPHALANTHERA

***Cephalanthera austinae* (A. Gray) A. Heller** (*Eburophyton austinae* [A. Gray] A. Heller)
Phantom orchid
Habitat/Range: Moist to mesic forests in the lowland zone; rare in the lower Fraser Valley, SE Vancouver Island (Saanich Peninsula) and the Gulf Islands; S to ID and CA.

COELOGLOSSUM

***Coeloglossum viride* (L.) C. Hartman ssp. *bracteatum* (Muhl. ex Willd.) Hult.** (*Habenaria viridis* [L.] R. Br. var. *bracteata* [Muhl. ex Willd.] A. Gray)
Long-bracted frog orchid
Habitat/Range: Moist to wet forests, grassy slopes, bogs and swamps in the lowland, steppe and montane zones; infrequent throughout BC east of the Coast-Cascade Mountains, rare along the coast; N to AK, YT and NT; E to NF and S to NC, OH, NE, NM and MT.

CORALLORHIZA

1. Tepals with purple stripes .... *C. striata*
1. Tepals not striped.
    2. Lateral sepals 1-nerved; lips 3–5 mm long; plants yellow or yellowish-green, usually less than 15 cm tall .... *C. trifida*
    2. Lateral sepals 3-nerved; lips 5–10 mm long; plants purple to brownish-red (rarely yellow), usually more than 15 cm tall .... *C. maculata*

Corallorhiza maculata **Raf.** (*C. mertensiana* Bong. = ssp. *mertensiana*)

Spotted coralroot (ssp. *maculata*), western coralroot (ssp. *mertensiana*)

Habitat/Range: Moist to dry forests in the lowland, steppe vegetation and montane zones; common throughout BC, especially along the coast, south of 55°N, rare northward; ssp. *maculata* - N to AK, E to NF and S to NC, TX, NM, MX and CA; Guatemala; ssp. *mertensiana* - N to SE AK and S to ID, WY and CA.

Notes: Two subspecies, with overlapping ranges, occur in BC.

1. Lips usually reddish-purple, not spotted, scarcely 3-lobed; spurs free, at least below the middle and conspicuously exserted .... ssp. *mertensiana* (Bong.) Calder & Taylor
1. Lips usually white, spotted with purple, deeply 3-lobed; spurs inconspicuous, reduced to a mere swelling .... ssp. *maculata*

Corallorhiza striata **Lindl.**

Striped coralroot

Habitat/Range: Moist forests in the lowland, steppe and montane zones; frequent throughout BC south of 56°N; E to PQ and S to TX, MX and CA.

Corallorhiza trifida **Chatelain**

Yellow coralroot

Habitat/Range: Moist coniferous forests in the steppe and montane zones; common throughout BC east of the Coast-Cascade Mountains, circumboreal, N to AK, E to NF and S to GA, CO and OR; Greenland and Eurasia.

CYPRIPEDIUM

1. Lips yellow, often veined or spotted with purple on the inner surface .... *C. calceolus*
1. Lips white or pinkish, tinged or veined with purple.
    2. Sepals shorter than the lips .... *C. passerinum*
    2. Sepals longer than the lips .... *C. montanum*

Cypripedium calceolus **L. ssp. *parviflorum* (Salisb.) Hult.**

Yellow ladyslipper

Habitat/Range: Moist streambanks and mossy bogs and swamps (often in calcareous sites) in the steppe and montane zones; infrequent E of the Coast-Cascade Mountains; circumpolar, N to AK, YT and NT, E to NF and S to GA, TX, AZ and OR; Eurasia.

Cypripedium montanum **Dougl. ex Lindl.**

Mountain ladyslipper

Habitat/Range: Moist to dry open woods in the steppe, montane and subalpine zones; infrequent E of the Coast-Cascade Mountains, rare northward; N to AK, E to SK and S to MT, ID and CA.

Cypripedium passerinum **Richards. var. *passerinum***

Sparrow's-egg ladyslipper

Habitat/Range: Moist woods, fens, edges of lakes and streams and talus slopes in the montane and subalpine zones; infrequent in N and E BC; N to AK, YT and NT and E to PQ.

EPIPACTIS

1. Lips 3-lobed; flowers large, sepals 12–16 mm long .... *E. gigantea*
1. Lips unlobed; flowers small, sepals less than 12 mm long.. *E. helleborine*

Epipactis gigantea **Dougl. ex Hook.**
Giant helleborine
Habitat/Range: Moist streambanks, marshes and swamps and especially around hot springs or calcareous seeps in the lowland, steppe and montane zones; rare in S BC; S to SD, TX and MX.

Epipactis helleborine **(L.) Crantz**
Helleborine
Habitat/Range: Moist to dry waste places, forests and fields in the lowland zone; rare garden escape in SW BC; introduced from Europe.

GOODYERA

1. Plants usually more than 20 cm tall; leaves 4–10 cm long; petals more than 4 mm long .... *G. oblongifolia*
1. Plants usually less than 15 cm tall; leaves less than 4 cm long; petals less than 4 mm long .... *G. repens*

Goodyera oblongifolia **Raf.**
Rattlesnake-plantain, or large-leaved rattlesnake orchid
Habitat/Range: Dry to moist, mossy forests in the lowland, steppe and montane zones; common throughout BC south of 56°N; N to AK, E to NS and S to ME, MI, WI, MN and MX.

Goodyera repens **(L.) R. Br. in Ait.**
Dwarf rattlesnake orchid
Habitat/Range: Dry to moist, mossy forests in the steppe and montane zones; frequent in BC east of the Coast-Cascade Mountains and S of 56°N; circumpolar, N to AK and YT, E to NF and S to NC, NV, SD and NM; Eurasia.

HAMMARBYA

Hammarbya paludosa **(L.) Kuntze** (*Malaxis paludosa* [L.] Sw.)
Bog adder's-mouth orchid
Habitat/Range: Wet bogs and muskegs in the lowland, steppe and montane zones; rare in BC north of 54°N; circumboreal, N to AK and NT, E to ON and S to MN; Eurasia.

LIPARIS

Liparis loeselii **(L.) Rich.**
Loesel's liparis
Habitat/Range: Moist thickets and bogs in the montane zone; rare in SC BC (known only from Mara Meadows Ecological Reserve); E to NS, NB and PE and S to AL, OH and NE.

LISTERA

1. Lips deeply cleft into two linear acute lobes .... *L. cordata*
1. Lips entire or shallowly cleft into obtuse lobes.
    2. Lips rectangular or oblong, broadly auriculated at base, without a pair of lateral teeth .... *L. borealis*
    2. Lips wedge-shaped, with a pair of lateral teeth, not auriculated at the base.

3. Ovaries glabrous; lips with a long slender tooth on each side at the base, not ciliate-margined .... *L. caurina*

3. Ovaries glandular; lips with a short triangular tooth on each side at the base, finely ciliate-margined .... *L. convallarioides*

Listera borealis **Morong**

Northern twayblade

Habitat/Range: Moist coniferous forests, edges of streams and meadows in the montane to alpine zones; infrequent throughout BC, except absent from the coast; N to AK, YT and NT, E to PQ and S to CO and WA.

Listera caurina **Piper** (*L. banksiana* Lindl.)

Northwestern twayblade

Habitat/Range: Moist forests, bogs and meadows in the montane and subalpine zone; infrequent in BC south of 55°N except frequent on the Queen Charlotte Islands; N to AK, E to AB and S to MT, ID and N CA.

Listera convallarioides **(Sw.) Nutt.**

Broad-leaved twayblade

Habitat/Range: Moist forests, bogs and streambanks in the montane zone; infrequent in BC south of 55°N; amphiberingian, N to AK, E to NF and S to NY, MI, CO, AZ and CA; Asia.

Listera cordata **(L.) R.Br. in Ait.**

Heart-leaved twayblade

Habitat/Range: Dry to moist forests, bogs and thickets in the montane zone; common throughout BC, except infrequent north of 56°N; circumboreal, N to AK, YT and NT, E to NF and S to NC, MI, MN and CA; Eurasia.

MALAXIS

Malaxis monophyllos **(L.) Sw.**

One-leaved malaxis

Habitat/Range: Moist forests and streambanks in the montane zone; var. *brachypoda* rare in coastal and N BC; var. *diphyllos* rare, known only from the Queen Charlotte Islands; amphiberingian, N to AK and YT, E to NF and S to NJ, PA, CO and CA; Eurasia, Japan.

Notes: Two varieties occur in BC.

1. Basal leaves one; flowers with lips pointed downward ....
.... var. *brachypoda* (Gray) Morris & Ames

1. Basal leaves two; flowers turned, with lips pointed upward .... var. *diphyllos* (Cham.) Luer

PIPERIA[12]

1. Spurs at most slightly longer than lips; racemes loosely flowered, usually not over 1 cm thick .... *P. unalascensis*
1. Spurs at least twice as long as lips; racemes usually at least 1.5 cm thick.
    2. Flowers greenish; sepals 3.5–4.5 mm long; racemes well over 10 cm long; plants generally distributed .... *P. elegans*
    2. Flowers white; sepals 5.0–5.5 mm long; racemes less than 10 cm long; plants coastal .... *P. maritima*

***Piperia elegans* (Lindl.) Rydb.** (*Habenaria elegans* [Lindl.] Boland.; *H. unalascensis* [Spreng.] S. Wats. var. *elata* [Jeps.] Correll; *Platanthera unalascensis* [Spreng.] Kurtz ssp. *elata* [Jeps.] Taylor & MacBryde)
Elegant rein orchid
Habitat/Range: Dry to moist forests in the lowland to montane zones; infrequent in S BC, south of 51°N; S to ID, MT and CA.

***Piperia maritima* Rydb.** (*Habenaria greenei* Jeps.; *H. unalascensis* [Spreng.] S. Wats. var. *maritima* [Greene] Correll; *Platanthera unalascensis* [Spreng.] Rydb. ssp. *maritima* [Rydb.] deFilipps)
Seaside rein orchid
Habitat/Range: Dry to vernally moist slopes and cliffs in the lowland zone; rare in coastal BC; S to CA.

***Piperia unalascensis* (Spreng.) Rydb.** (*Habenaria unalascensis* [Spreng.] S. Wats.; *Platanthera unalascensis* [Spreng.] Kurtz)
Alaska rein orchid
Habitat/Range: Dry to moist coniferous forests, grassy slopes and streambanks in the lowland, steppe to lower subalpine zones; common throughout BC south of 56°N, especially east of the Coast-Cascade Mountains; N to AK, E to PQ and S to SD, CO and CA; Baja CA.
Notes: A poorly defined species (at least in the herbarium), *Piperia candida*, has been named recently and attributed to our flora (see Morgan and Ackerman 1990). Until further study, it has not been included here.

PLATANTHERA[13]

1. Leaves one or two, essentially basal; stems lacking bracts, or with one at or above the middle of the stems.
    2. Leaves one .... *P. obtusata*
    2. Leaves two.
        3. Plants less than 20 cm tall; leaves 2.5–6 cm long, erect or spreading; spurs about 1 mm long .... *P. chorisiana*
        3. Plants 20–60 cm tall; leaves 6–16 cm long, lying flat on ground; spurs 15–25 mm long .... *P. orbiculata*
1. Leaves and bracts usually several, cauline or sub-basal.
    4. Flowers usually white, rarely green, fragrant; lips abruptly dilated toward base .... *P. dilatata*
    4. Flowers usually greenish, not fragrant; lips gradually tapered toward base.

---

12 Often included in *Habenaria* or *Platanthera*.
13 A difficult genus, with many taxonomic interpretations. Often placed in *Habenaria*.

5. Spurs saccate; lips linear to lanceolate; racemes elongate, flowers scattered .... *P. stricta*
5. Spurs cylindrical; lips elliptic to ovate-lanceolate; racemes congested .... *P. hyperborea*

***Platanthera chorisiana* (Cham.) Reichenb.** (*Habenaria chorisiana* Cham.)
Small bog orchid, or chamisso's rein orchid
Habitat/Range: Sphagnum bogs and swamps in the lowland zone; rare in S coastal BC, infrequent northward in coastal BC; N to AK and S to WA; Japan.

***Platanthera dilatata* (Pursh) Lindl. ex Beck** (*Habenaria dilatata* [Pursh] Hook. = var. *dilatata*; *H. dilatata* var. *albiflora* [Cham.] Correll = var. *albiflora*; *H. dilatata* var. *leucostachys* [Lindl.] Ames = var. *leucostachys*)
White bog orchid, or fragrant white rein orchid
Habitat/Range: Wet meadows, seepage slopes, bogs, stream and lake margins and open forests in the montane to alpine zones; common throughout BC, except var. *albiflora*, rare in S BC; N to AK, YT and NT, E to NF and S to PA, MI, WI, SD, CO and CA; Greenland.
Notes: Three varieties occur in BC.

1. Spurs about twice as long as lips .... var. *leucostachys* (Lindl.) Leur
1. Spurs shorter than lips or only slightly longer.
   2. Spurs up to two-thirds the length of lips .... var. *albiflora* (Cham.) Ledeb.
   2. Spurs equalling the lips or slightly longer .... var. *dilatata*

***Platanthera hyperborea* (L.) Lindl.** (*Habenaria hyperborea* [L.] R. Br.)
Green-flowered bog orchid, or green-flowered rein orchid
Habitat/Range: Wet grassy slopes, stream and lake margins, bogs, swamps and open forests in the montane to alpine zones; common throughout BC; circumpolar, N to AK, YT and NT, E to NF and S to PA, NE, NM, MX and CA; Greenland, Iceland, Asia.

***Platanthera obtusata* (Banks ex Pursh) Lindl.** (*Habenaria obtusata* [Banks ex Pursh] Richards.)
One-leaved rein orchid
Habitat/Range: Moist to wet forests, bogs, swamps and streambanks in the montane zone; infrequent in S BC, more frequent northward, except absent from the coast; N to AK, E to NF and S to NY, MI, MN and CO; Norway.

***Platanthera orbiculata* (Pursh) Lindl.** (*Habenaria orbiculata* [Pursh] Torr.)
Large round-leaved rein orchid
Habitat/Range: Dry to moist forests, bogs and swamps in the montane zone; infrequent in BC south of 56°N, except rare on the coast; N to AK, E to Labr. and S to NY, OH, IN, MN, MT and WA.

***Platanthera stricta* Lindl.** (*Habenaria stricta* [Lindl.] Rydb., *H. saccata* Greene)
Slender bog orchid, or rein orchid
Habitat/Range: Moist meadows, coniferous forests, swamps and bogs in the lowland, montane and lower subalpine zones; common throughout BC south of 55°N, rare northward; N to AK, E to AB and S to WY, CO, NM and CA.

SPIRANTHES

***Spiranthes romanzoffiana* Cham.**
Hooded ladies' tresses
Habitat/Range: Dry to moist sites in a wide variety of non-forested habitats in the lowland to montane zones; common throughout BC, south of 55°N, infrequent northward and on Vancouver Island; N to AK, YT, NT; E to NF and S to PA, MI, IA, SD, CO and CA; Ireland, Scotland.

POACEAE[14]

1. Spikelets dorsally compressed, the glumes and fertile lemma more or less flattened or merely arching, the spikelets sometimes terete, but then the caryopsis (i.e., fruit) somewhat plano-convex; perfect floret 1, terminal in the spikelets, usually subtended by a sterile or staminate floret; spikelets disarticulating below the glumes, the spikelets falling entire, either singly or in pairs with segments of the rachis .... GROUP 1

1. Spikelets usually more or less laterally compressed, sometimes terete; perfect florets 1 to many, the reduced florets, when present, above the perfect ones (except in *Arrhenatherum, Hierochloë* and *Phalaris* - the first has 1 subtending staminate floret while the latter 2 have 2 subtending rudimentary or staminate florets); spikelets disarticulating either above or below the glumes.

    2. Glumes lacking; spikelets 1-flowered; palea subequal to the lemma and 3-nerved.

        3. Plants perennials, aquatic; lemmas 5-nerved, about 5 mm long, bristly-pubescent, unawned .... *Leersia*

        3. Plants annuals, not aquatic; lemmas 1-nerved, about 1 mm long, bristly-pubescent only on the keels, with awn-like tips .... *Coleanthus*

    2. Glumes present (only 1 in *Lolium*, except on the terminal spikelet); spikelets various; palea, if present, 2-nerved.

        4. Spikelets usually with 2 staminate or rudimentary florets (which may look like hairs or scales) below the terminal fertile floret.

            5. Glumes distinctly unequal, the first 1-nerved and only about half as long as the second; lower two florets sterile, the lemmas awned from the back; lodicules lacking .... *Anthoxanthum*

            5. Glumes subequal, both usually 3-nerved; lower two florets either staminate or sterile, the lemmas awnless; lodicules usually present.

                6. Lower florets staminate; spikelets brown and shiny; auricles absent ...... *Hierochloë*

                6. Lower florets empty and much reduced; spikelets green or yellowish; auricles present .... *Phalaris*

        4. Spikelets without a reduced floret below the perfect floret(s).

            7. Spikelets sessile or subsessile (when more than one at a node some of the group usually pedicellate); inflorescences single spikes, or 2 or more digitately or racemosely arranged spicate branches (the spicate branches of *Beckmannia* sometimes arranged in panicles) .... GROUP II

            7. Spikelets pedicellate (if sessile or subsessile then inflorescences not as above); inflorescences open or contracted, sometimes spike-like panicles, rarely racemes, usually reduced to a single spikelet in *Danthonia unispicata*.

                8. Spikelets 2- to many-flowered.

                    9. Glumes shorter than the first floret; lemmas awnless or awned from the tip or from a bifid apex, the awns rarely slightly dorsal .... GROUP III

                    9. Glumes, at least the second glume, as long or longer than the lowest floret, often as long as the spikelet (sometimes slightly shorter than the first floret in *Koeleria* and *Sphenopholis*); lemmas with dorsal awns or awnless (awned from a bifid apex in *Danthonia* and *Koeleria*) .... GROUP IV

---

14 Key adapted from Holmgren and Holmgren (1977).

8. Spikelets 1-flowered (rarely 2 in *Calamagrostis*) .... GROUP V

GROUP I

1. Glumes indurate; sterile lemma, fertile lemma, and palea hyaline or membranous, the 2 lemmas usually alike in texture .... *Schizachyrium*
1. Glumes membranous; sterile lemma resembling the second glume and appearing like a third glume; fertile lemma and palea indurate or at least firmer than the glumes.
    2. Spikelets subtended by bristles or enclosed in prickly burs (involucres).
        3. Spikelets subtended by 1-many bristles, the spikelets deciduous and the bristles persistent .... *Setaria*
        3. Spikelets enclosed in prickly burs, the burs dropping intact from the rachis at maturity .... *Cenchrus*
    2. Spikelets not subtended by bristles or enclosed in prickly burs.
        4. Glumes and sterile lemma usually awned; spikelets in irregular clusters along one side of the panicle branches, the spikelets stiffly hispid .... *Echinochloa*
        4. Glumes and sterile lemma awnless.
            5. Spikelets in 1-sided spikes or racemes .... *Digitaria*
            5. Spikelets in open or compact panicles .... *Panicum*

GROUP II

1. Spikelets or spikelet clusters alternate on opposite sides of the rachis; spikes terminal, solitary; leaves often auriculate.
    2. First glume lacking, except in the terminal spikelet; spikelets with the backs of lemmas against the rachis .... *Lolium*
    2. Both glumes present on all spikelets; spikelets with the edges of the lemmas against the rachis.
        3. Spikelets in clusters of three at each node, each spikelet with only 1 floret, those of the lateral spikelet usually sterile or staminate, often highly reduced .... *Hordeum*
        3. Spikelets 1–4 (5) at each node, each with 2 or more florets, the terminal florets often reduced, spikelets all alike.
            4. Plants annual; weedy or cultivated as cereal grains or planted for erosion control; introduced.
                5. Spikelets paired at each node; glumes strongly keeled below .... *Secale*
                5. Spikelets solitary at each node; glumes rounded below .... *Triticum*
            4. Plants perennial; mostly not weedy, a few used in erosion control or range "improvement"; mostly native species.
                6. Rachis internodes 0.5–3 mm long; spikelets strongly diverging from the rachis .... *Agropyron*
                6. Rachis internodes 5–25 mm long; spikelets ascending or appressed to the rachis.
                    7. Plants sterile hybrids; anthers not-dehiscing even at maturity, remaining attached .... *x Elymordeum*

7. Plants fertile; anthers dehiscing at maturity, then falling off.

8. Spikelets always 2 or more at each node; lemma awns absent or less than 3.5 mm long; anthers more than 2.5 mm long; plants rhizomatous or glumes both short and awnlike .... *Leymus*

8. Spikelets 1–4 (5) at each node; lemmas often with an awn greater than 3 mm; anther length varied; plants caespitose or the glumes not awnlike (or both) .... *Elymus*

1. Spikelets in 2 rows, along 2 sides of a 3-angled, continuous rachis, forming 1-sided spikes or racemes, these spicate branches usually 2 to several, digitately, racemosely or paniculately arranged on a central axis; leaves non-auriculate.

9. Spikelets with 2–3 perfect florets .... *Leptochloa*

9. Spikelets with only one perfect floret, often with additional imperfect florets, these usually rudimentary.

10. Spikelets with 1 or more modified florets about the perfect one .... *Bouteloua*

10. Spikelets without additional modified florets, the rachilla sometimes prolonged.

11. Spikelets disarticulating above the glumes .... *Cynodon*

11. Spikelets disarticulating below the glumes and falling entirely.

12. Spikes (2) 4–15 cm long; glumes narrow-lanceolate to subulate, 3.5–20 mm long; lemma 6.2–9.5 mm long .... *Spartina*

12. Spikes less than 1.8 cm long; glumes broad, rounded-triangular, 2–3 mm long; lemma 1.8–3.5 mm long .... *Beckmannia*

GROUP III

1. Plants dioecious or monoecious; lemmas glabrous to scabrous, not villous .... *Distichlis*

1. Plants not dioecious, except in some species of *Poa* with villous lemmas.

2. Plants tall, stout reeds, 2–3 m tall, with large, plume-like panicles; rachillas with long, spreading hairs .... *Phragmites*

2. Plants not reed-like, rarely up to 2 m tall; rachillas without long, spreading hairs.

3. Lemmas mostly 5- or more-nerved, if 3-nerved the nerves usually obscure.

4. Callus of the florets bearded.

5. Lemmas awnless.

6. Plants aquatic; leaf blades flat at the tips .... *Arctophila*

6. Plants terrestrial; leaf blades usually curved and prow-like at the tips .... *Poa*

5. Lemmas awned.

7. Lemmas awned from a bifid apex; sheaths closed .... *Schizachne*

7. Lemmas awned, the first from the tip, the second deeply bifid and awned from the back; sheaths open .... *Ventenata*

4. Callus of the florets not bearded.

8. Lemmas keeled on the back.

9. Spikelets strongly compressed, crowded in dense, 1-sided panicles ... *Dactylis*

9. Spikelets not strongly compressed (except some *Bromus*), not crowded in dense 1-sided panicles.

10. Spikelets large, mostly more than 15 mm long; lemmas awned, often from a minutely bifid apex, or nearly awnless .... *Bromus*

10. Spikelets small, mostly less than 10 mm long; lemmas awnless .... *Poa*

8. Lemmas rounded on the back.

11. Spikelets dimorphic, borne mostly in pairs, one of these fertile ...... *Cynosurus*

11. Spikelets all alike or at least not dimorphic and in part sterile.

12. Glumes papery; spikelets disarticulating above or below the glumes .... *Melica*

12. Glumes not papery; spikelets disarticulating above the glumes.

13. Nerves of the lemma parallel, not converging at the summit, or only slightly so.

14. Inflorescences simple 1-sided racemes; lemmas awned .... *Pleuropogon*

14. Inflorescences open to contracted panicles; lemmas awnless.

15. Nerves of the lemmas obscure, usually 5 ...... *Puccinellia*

15. Nerves of the lemmas usually prominent, 5–7.

16. Leaf sheaths open, their margins free and overlapping .... *Torreyochloa*

16. Leaf sheaths (at least the upper ones) closed, or nearly so .... *Glyceria*

13. Nerves of the lemma converging toward the summit.

17. Lemmas tapered to a point, usually awned or sometimes awn-tipped (this appearance often exaggerated by involution of the lemma); leaf sheaths open.

18. Plants annual .... *Vulpia*

18. Plants perennial .... *Festuca*

17. Lemmas broader, or if somewhat tapered then awned from a bifid apex; leaf sheaths closed, at least in the lower half.

19. Spikelets large, mostly more than 15 mm long; lemmas often awned from a bifid apex, or nearly awnless .... *Bromus*

19. Spikelets small, mostly less than 10 mm long; lemmas awnless .... *Poa*

3. Lemmas 3-nerved, the nerves usually prominent.

20. Spikelets (5–) 6- to many-flowered .... *Eragrostis*

20. Spikelets 2- or 3-flowered .... *Muhlenbergia*

GROUP IV

1. Lemmas awned from the apex or awnless.
    2. Lemmas awned.
        3. Lemma awns arising between 2 lobes .... *Danthonia*
        3. Lemma awns arising nearly imperceptibly from a narrow, acuminate, unlobed apex ... *Vulpia*
    2. Lemmas awnless or very shortly awned from a minutely bifid apex.
        4. Lemmas nearly as wide as long .... *Briza*
        4. Lemmas much longer than wide.
            5. Plants strongly rhizomatous perennials; spikelets in open panicles .... *Scolochloa*
            5. Plants strongly caespitose perennials; spikelets in open or congested spikelike panicles.
                6. Sheaths hispid; upper glume widest above middle .... *Koeleria*
                6. Sheaths glabrous; upper glume widest below middle .... *Sphenopholus*
1. Lemmas (at least the uppermost) awned dorsally.
    7. Spikelets with 2 or more perfect florets, the reduced florets, when present, terminal in the spikelets.
        8. Spikelets small, the glumes 2–9 mm long.
            9. Lemma awns attached above the middle; lemmas (3.5) 4–9 mm long .... *Trisetum*
            9. Lemma awns attached below the middle; lemmas 1.5–4 (4.5) mm long.
                10. Plants delicate annuals; rachillas not prolonged beyond the second floret .... *Aira*
                10. Plants perennials; rachillas often prolonged beyond the second floret.
                    11. Leaf blades flat, 3–6 mm wide, rolled in the bud; ligules obtuse to truncate; spikelets purplish .... *Vahlodea*
                    11. Leaf blades usually folded or involute, 1–3 (4) mm wide, folded in the bud; ligules acute or acuminate; spikelets greenish or tawny .... *Deschampsia*
        8. Spikelets large, the glumes 8–30 mm long.
            12. Plants large annuals, (2.5) 4–19 dm tall; glumes 17–30 mm long; panicles open, drooping, 10–38 cm long .... *Avena*
            12. Plants small perennials, 0.8–1.8 (2) dm tall; glumes 8–11.5 mm long; panicles narrow, erect, 3–5 (7.5) cm long .... *Helictotrichon*
    7. Spikelets with 1 perfect floret and 1 staminate floret.
        13. Lower floret staminate and awned, the awn twisted, geniculate, exserted; upper floret perfect, awnless or with a small, hair-like, nearly terminal awn .... *Arrhenatherum*
        13. Lower floret perfect, awnless; upper floret staminate, with a hooked, curved or bent awn .... *Holcus*

GROUP V

1. Lemmas thickly membranous to indurate, terminally awned.
   2. Awns 3-branched, junction of awn and lemma not evident .... *Aristida*
   2. Awns unbranched, junction of awn and lemma evident.
      3. Glumes narrowing in the distal third, acute; florets usually less than 5 times as long as wide, never pilose .... *Oryzopsis*
      3. Glumes tapering from midlength or below, acuminate to acute; lemmas usually more than 5 times as long as wide, pilose in *S. hymenoides*.
         4. Lemma margins strongly overlapping, difficult to pry apart; palea glabrous, less than 1/3 the length of the lemma .... *Nassella*
         4. Lemma margins slightly overlapping, sometimes separate at maturity, easily pried apart; palea usually pubescent, more than 1/3 the length of the lemma .... *Stipa*
1. Lemmas membraneous, not evidently indurate, awned or awnless.
   5. Glumes, at least the second, equalling or usually longer than the floret.
      6. Spikelets disarticulating above the glumes.
         7. Glumes abruptly short-awned from an obtuse apex, ciliate on the prominent keels .... *Phleum*
         7. Glumes not abruptly awned, if awned at all, not ciliate on the keels.
            8. Lemmas 1-nerved, awnless .... *Sporobolus*
            8. Lemmas 3- to 5-nerved, awned or sometimes awnless.
               9. Lemmas awned from an entire or minutely cleft apex, the body of the lemmas 3-nerved .... *Muhlenbergia*
               9. Lemmas with dorsal awns, or if awnless, 5-nerved.
                  10. Lemma awns evident, more than 5.5 mm long, arising from just below the apex of the lemmas; lemmas firm to subindurate .... *Apera*
                  10. Lemma awns less than 4.5 mm long or absent, or if longer, arising at or below the midlength of the lemmas (in *Calamagrostis purpurascens*); lemmas membranous.
                      11. Callus of the floret evidently bearded .... *Calamagrostis*
                      11. Callus of the floret glabrous to minutely bearded.
                          12. Glumes 3-nerved; palea usually well developed ... *Arctagrostis*
                          12. Glumes 5-nerved (sometimes 3-); palea often absent or at least poorly developed .... *Agrostis*
      6. Spikelets disarticulating below the glumes, the spikelets falling entire.
         13. Glumes awned.
             14. Lemmas 0.7–1.5 mm long; anthers small, 0.4–0.7 mm long .... *Polypogon*
             14. Lemmas (1.2) 1.7–2.5 mm long; anthers 1.1–2.3 mm long .... *Phleum*

13. Glumes awnless.

15. Lemmas with a short awn from the notched apex; panicles usually somewhat open .... *Cinna*

15. Lemmas awned from the middle or below; panicles dense, cylindrical, spike-like .... *Alopecurus*

5. Glumes shorter than the floret.

16. Glumes or lemmas awned .... *Muhlenbergia*

16. Glumes or lemmas awnless.

17. Spikelets disarticulating below the glumes.

18. Glumes dissimilar, the first linear, the second broadly oblanceolate to obovate; ligules 1.5–3.5 mm long .... *Sphenopholis*

18. Glumes subequal, semicircular; ligules 5–8.5 mm long .... *Beckmannia*

17. Spikelets disarticulating above the glumes.

19. Callus of the florets bearded; spikelets 6.5 mm or more long.

20. Panicles spikelike; rachillas prolonged; plants of coastal SW BC .... *Ammophila*

20. Panicles open; rachillas not prolonged; plants of SE BC .... *Calamovilfa*

19. Callus of the florets glabrous to scabrous; spikelets 1–4.5 (5.5) mm long.

21. Lemmas 1-nerved; fruit not a true caryopsis as the ovary wall and seed coat are not fused .... *Sporobolus*

21. Lemmas 3-nerved; fruit a true caryopsis.

22. Sheath margins free to the base; nerves of the lemmas converging towards the apex .... *Muhlenbergia*

22. Sheath margins united to above the middle or near the ligule; nerves of the lemmas parallel, not converging towards the apex.

23. Spikelets 1.2–1.7 mm long; ligules 1–1.5 mm long, glabrous .... *Phippsia*[15]

23. Spikelets 3–4 mm long; ligules 2–8 mm long, erose to almost entire .... *Catabrosa*

AGROPYRON[16,17]

1. Spikelets spreading at an angle of 30° or more from the rachis; distance between spikelets less than 1 mm .... *A. cristatum*

1. Spikelets forming an angle of less than 30° with the rachis; distance between spikelets more than 1 mm.

---

15 *Phippsia algida* (Soland.) R. Br., known from moist, alpine gravelly sites in extreme SW YT, has yet to be collected in BC but should be expected.

16 Contributed by M.E. Barkworth.

17 Other species previously included in *Agropyron* now appear in *Elymus* or *Leymus*.

2. Lemmas awnless .... *A. fragile*

2. Lemmas shortly awned .... *A. desertorum*

***Agropyron cristatum* (L.) Gaertn.** (*A. pectiniforme* Roem. & Schult.)
Crested wheatgrass
Habitat/Range: Roadsides, fields and waste places; frequent in SC and SE BC, rare in SW BC; introduced from Eurasia.
Notes: The three *Agropyron* species occurring in BC hybridize readily when growing together, forming morphologically intermediate plants. In addition, some cultivars are also intermediate, being derived from hybrids. Consequently, identification of *Agropyron* species in North America is often extremely difficult. In many instances, it may be best to refer to plants of this genus as *A. cristatum sens. lat.*, the last phrase indicating no distinction is being made between the three species.

***Agropyron desertorum* (Fisch. ex Link) Schult.** (*A. sibiricum* [Willd.] Beauv.)
Desert wheatgrass
Habitat/Range: Roadsides, fields and waste places; rare in BC east of of the Coast-Cascade Mountains and S of 56°N; introduced from Eurasia.

***Agropyron fragile* Roth** (*A. sibiricum* [Willd.] Beauv.)
Siberian wheatgrass
Habitat/Range: Roadsides, fields and waste places; infrequent in SC and SE BC; introduced from Eurasia.

AGROSTIS[18]

1. Stolons (above-ground horizontal stems) or rhizomes (below-ground horizontal stems) present, over 2 cm long.

2. Stolons (above-ground horizontal stems) present; rhizomes (below-ground horizontal stems) absent .... *A. stolonifera*

2. Stolons absent; rhizomes present.

3. Palea up to 0.2 mm long or absent .... *A. pallens*

3. Palea more than 0.6 mm long (at least half as long as the lemmas).

4. Ligules 2–6 mm long, longer than wide; rhizomes abundant, often more than 10 cm long; inflorescence branches densely covered in spikelets to the bases (except in shade plants) .... *A. gigantea*

4. Ligules up to 2 mm long, wider than long; rhizomes sparse, usually less than 5 mm long; inflorescence branches with well-spaced spikelets, none or only a few of these near the bases .... *A. capillaris*

1. Rhizomes or stolons absent, or if present less than 2 cm long.

5. Mature inflorescences more or less open with the branches spreading and visible.

6. Palea present, at least 1/2 as long as the lemmas.

7. Rachilla present behind the palea as a minute bristle, at least in some spikelets; usually in undisturbed native habitats.

---

18 Contributed by M.J. Harvey.

8. Anthers 0.8–1.5 mm long; plants of the coastal lowland and montane zones .... *A. aequivalvis*

8. Anthers 0.6–0.8 mm long; plants of the subalpine or alpine zones .... *A. humilis*

7. Rachilla absent; usually in disturbed habitats.

9. Ligules 2–6 mm long, longer than wide; rhizomes abundant, greater than 5 mm long; inflorescence branches densely covered in spikelets to the bases (except in shade plants) .... *A. gigantea*

9. Ligules up to 2 mm long, wider than long; rhizomes sparse, usually less than 5 mm long; inflorescence branches with well-spaced spikelets, none or only a few of these near the bases .... *A. capillaris*

6. Palea absent, or if present then less than 1/4 as long as the lemmas.

10. Lemmas with awns arising from the backs.

11. Mature inflorescences about as wide as long, the branches capillary, diffuse and lax .... *A. scabra*

11. Mature inflorescences at least twice as long as wide, branches rather stiff, more or less ascending .... *A. mertensii*

10. Lemmas without awns.

12. Mature inflorescences about as wide as long, usually over 10 cm long, often large, upright, the branches capillary, diffuse without spikelets near the bases of the branches .... *A. scabra*

12. Mature inflorescences at least twice as long as wide, if over 10 cm long then either narrow, lax and drooping to one side or with spikelets crowded to the bases of the branches.

13. Anthers 1.0–1.7 mm long .... *A. pallens*

13. Anthers usually less than 1 mm long.

14. Plants 12–35 cm tall; inflorescences usually less than 10 cm long; plants of subalpine and alpine zones, occasionally in disturbed sites in the montane zone .... *A. idahoensis*

14. Plants usually over 35 cm tall; infloresences usually over 10 cm long; plants of the lowland zone .... *A. oregonensis*

5. Mature inflorescences more or less constricted with the branches mainly appressed to the central axis.

15. Palea present, at least 1/2 as long as the lemma.

16. Anthers more than 1 mm long; plants usually vigorously spreading, in disturbed sites from the lowland to subalpine zones .... *A. stolonifera*

16. Anthers less than 0.8 mm long; plants usually tufted, in natural sites in the subalpine and alpine zones .... *A. humilis*

15. Palea absent, or if present, then only as a minute membrane.

17. Lemmas awnless.

18. Leaf blades less than 2 mm wide; inflorescences 2–8 cm long with few or no spikelets near the branch bases; plants of the subalpine and alpine zones .... *A. variabilis*

18. Leaf blades 2–8 mm wide; inflorescences 4.5–18 cm long with the branches densely covered in spikelets to the bases; plants of the lowland to subalpine zones .... *A. exarata*

17. Lemmas with awns arising from the backs.

19. Spikelets usually 3–4 mm long; all shoots of the plants flowering; plants of dry sites in the lowland zone of S Vancouver Island and the Gulf Islands .... *A. microphylla*

19. Spikelets usually 1.8–3 mm long; some shoots of the plants nonflowering; plants of various sites and zones but not as above.

20. Inflorescences dense, the branches barely visible, covered with spikelets to the base .... *A. exarata*

20. Inflorescences loose, the branches visible with fewer spikelets borne mostly above mid-point .... *A. mertensii*

***Agrostis aequivalvis* (Trin.) Trin.** (*Agrostis thurberiana* auct. non A.S. Hitchc., *Podagrostis aequivalvis* [Trin.] Scribn. & Merr.)
Alaska bentgrass
Habitat/Range: Moist stream and bog margins in the lowland and lower montane zones; common on Vancouver Island, the Queen Charlotte Islands and adjacent coast; N to AK and S to OR.

***Agrostis capillaris* L.** (*A. tenuis* Sibth. non Vasey)
Colonial bentgrass
Habitat/Range: Lawns, fields, roadsides and meadows in the lowland zone; common on S Vancouver Island and the adjacent mainland; introduced from Europe.

***Agrostis exarata* Trin.** (*A. exarata* vars. *minor* Hook. and *monolepis* [Torr.] A.S. Hitchc.)
Spike bentgrass
Habitat/Range: Mesic to wet fields, tidal marshes, roadsides and disturbed sites in the lowland, steppe and montane zones; common throughout BC; amphiberingian, N to AK, YT and NT, E to S SK, and S to SD, NE, TX, NM, N MX and CA; E Asia, S. America.
Notes: *A. exarata* exhibits marked variation due to genetics and environment. Awned and awnless spikelets may occur on separate plants or in the same inflorescence. Plants also vary greatly in total height and size and colour of the inflorescence. For these reasons several varieties have been proposed including vars. *minor* and *monolepis.*

***Agrostis gigantea* Roth** (*A. alba* auct. non L., *A. stolonifera* var. *major* [Gaud.] Farw.)
Redtop
Habitat/Range: Dry roadsides, fields and waste places; common throughout BC; introduced from Europe.
Notes: This species was formerly combined with *A. stolonifera* under the illegitimate name *A. alba.* It is distinguished by its underground rhizomes, its open inflorescences and tolerance of much drier conditions.

***Agrostis humilis* Vasey** (*A. thurberiana* auct. non A.S. Hitchc., *Podagrostis humilis* [Vasey] Bjoerkm.)
Alpine bentgrass
Habitat/Range: Mesic to wet meadows, rocky slopes and moraines from the montane to alpine zones; common throughout BC; E to AB and S to CO, UT, NV and OR.
Notes: Plants with open inflorescences have usually been put into *A. thurberiana* but are here included in *A. humilis* because of the existence of intermediates. *A. variabilis*, another alpine species, closely resembles *A. humilis* and should be confirmed by its lack of a palea.

Agrostis idahoensis **Nash**
Idaho bentgrass
Habitat/Range: Mesic to wet bogs, meadows and seepage areas in the subalpine and alpine zones, occasionally in ditches and logging areas in the montane zone; infrequent on Vancouver Island, the adjacent mainland and N in the Coast Mountains to 54°N; N to AK, E to AB and S to CO, NM and CA.
Notes: This species is often confused with the larger panicled *A. scabra* and the awned *A. mertensii.*

Agrostis mertensii **Trin.** (*A. borealis* Hartm.)
Northern bentgrass
Habitat/Range: Mesic to wet meadows, streambanks, cliffs and seepage slopes in the lowland to alpine zones; infrequent throughout BC; circumboreal, N to AK, E to NF and S to NC, CO and WA; Greenland, N. Eurasia.

Agrostis microphylla **Steud.** (*A. inflata* Scribn.)
Small-leaved bentgrass
Habitat/Range: Wet vernal sites and mesic to dry grassy meadows and open forests in the lowland zone; locally frequent on SE Vancouver Island and the Gulf Islands; S to NW MX.

Agrostis oregonensis **Vasey**
Oregon bentgrass
Habitat/Range: Moist to wet meadows, streamsides, bogs, fens and forest openings in the lowland zone; common in coastal BC; N to AK, E to AB and S to NV and CA.

Agrostis pallens **Trin.** (*A. diegoensis* Vasey, *A. lepida* A.S. Hitchc.)
Dune bentgrass
Habitat/Range: Mesic to moist sand dunes and rocky seacliffs in the lowland zone; rare on the Queen Charlotte Islands and the W coast of Vancouver Island; S to CA.

Agrostis scabra **Willd.** (*A. hiemalis* auct. non [Walt.] BSP)
Hair or winter bentgrass
Habitat/Range: Moist to dry meadows, rock outcrops, forest openings and roadsides in the lowland to subalpine zones; common throughout BC; amphiberingian, N to AK, YT and NT, E to NF and S to FL, TX, MX and CA; E Asia.
Notes: *Agrostis perennans* (Walt.) Tuckerm., hitherto regarded as an eastern species, has recently been found in WA and can be expected in BC. It differs from *A. scabra* in its slightly more condensed but still capillary inflorescence and few or no basal leaves at flowering time but several well-developed stem leaves. It grows on moist to wet cliffs and streamsides. See also comments under *A. idahoensis.*

Agrostis stolonifera **L.** (*A. alba* auct. non L., *A. palustris* Huds.)
Creeping bentgrass
Habitat/Range: Mesic to wet lawns, fields, ditches, pond margins, lakeshores and saltmarshes in the lowland to subalpine zones; common throughout BC; introduced from Europe.
Notes: See comments under *A. gigantea.*

Agrostis variabilis **Rydb.**
Mountain bentgrass
Habitat/Range: Mesic to moist meadows and rocky slopes in the subalpine to alpine zones; common in BC in and E of the Coast-Cascade Mountains; N to AK, E to AB and S to CO and CA.
Notes: See comments under *A. humilis.*

AIRA

1. Panicles compact, more or less spikelike, 1–3 cm long; lemmas (2.5) 3–3.2 (3.5) mm long .... *A. praecox*
1. Panicles open, 2–6 cm long; lemmas 2–2.3 mm long .... *A. caryophyllea*

Aira caryophyllea **L.**
Silver hairgrass
Habitat/Range: Moist to dry gravelly or rocky sites; common on S Vancouver Island, less frequent in the Gulf Islands and lower mainland, rare in the Queen Charlotte Islands; introduced from Europe.

Aira praecox **L.**
Early hairgrass
Habitat/Range: Moist to dry gravelly or rocky sites; common on S Vancouver Island, the Gulf Islands and lower mainland, infrequent N to the Queen Charlotte Islands; introduced from Europe.

ALOPECURUS[19]

1. Spikelets over 4 mm long (excluding awns); anthers more than 2 mm long; glumes acute.
   2. Spikes oblong to short-cylindric; spikelets less than 4.5 mm long; glumes densely covered over entire surface with long woolly hairs; lemma apices obtuse .... *A. alpinus*
   2. Spikes long-cylindric; spikelets often over 4.5 mm long; glumes with long hairs mainly restricted to nerves and keel; lemma apices acute .... *A. pratensis*
1. Spikelets less than 4 mm long (excluding awns); anthers less than 2 mm long; glumes obtuse (rarely acute).
   3. Anthers 1 mm long or less; perennials or annuals.
      4. Plants perennials; awns very short, exserted less than 1.5 mm beyond glumes .... *A. aequalis*
      4. Plants annuals; awns extending 2.5 mm or more beyond glumes .... *A. carolinianus*
   3. Anthers more than 1 mm long; perennials.
      5. Spikes oblong to short-cylindric; spikelets more than 3 mm long; glumes densely covered over entire surface with long woolly hairs; lemma apices obtuse .... *A. alpinus*
      5. Spikes long-cylindric; spikelets 3.2 mm long or less; glumes with long hairs restricted to nerves and keel; lemma apices nearly truncate .... *A. geniculatus*

Alopecurus aequalis **Sobol.**
Little meadow-foxtail, or shortawn foxtail
Habitat/Range: Wet to mesic (including desiccating) shores, sloughs, and streambanks from the lowland to subalpine zones; common throughout BC; circumboreal, N to AK, YT and NT, E to Greenland and S to NJ, IN, KS, NM and CA; Eurasia, N Africa.

Alopecurus alpinus **J.E. Smith** (*A. alpinus* var. *glaucus* [Less.] Krylov)
Alpine meadow-foxtail
Habitat/Range: Moist subalpine meadows; rare in N BC, known only from Beatton River and Pink Mountain; circumpolar, N to AK, YT and NT, E to Greenland and S to CO and UT; Eurasia.

---

19 Contributed by W.J. Crins.

***Alopecurus carolinianus* Walt.** (*A. macounii* Vasey)
Carolina meadow-foxtail
Habitat/Range: Wet to moist (including desiccating) vernal pools, roadsides and meadows in the lowland zone; rare on S Vancouver Island and the lower Fraser River valley; E to S SK and CT and S to FL, TX, NM and CA.

***Alopecurus geniculatus* L.** (*A. pallescens* Piper)
Water meadow-foxtail
Habitat/Range: Wet to mesic shores, meadows, ditches and roadsides from the lowland to montane (rarely subalpine) zones; common in SW BC, north to the Queen Charlotte Islands, rare in SE BC, (known only from Nelson); introduced from Eurasia.
Notes: Although thought to be introduced, it occurs in natural habitats, especially in the Queen Charlotte Islands. Includes plants misidentified as *A. saccatus* Vasey, which is a distinct, annual species of CA and OR.

***Alopecurus pratensis* L.**
Meadow-foxtail
Habitat/Range: Wet to mesic meadows, fallow fields, and roadsides; infrequent in the lowland, steppe and montane zones in SC and SW BC; introduced from Eurasia.

AMMOPHILA

***Ammophila arenaria* (L.) Link**
European beachgrass
Habitat/Range: Sand dunes and sandy beaches; frequent on the Queen Charlotte Islands and Vancouver Island; introduced from Europe.

ANTHOXANTHUM

1. Plants annuals, mostly 1–3 dm tall; blades 1–2 mm wide; ligules 0.5–2 mm long .... *A. puelii*
1. Plants perennials, mostly 3–6 dm tall; blades (2) 3–7 mm wide; ligules (1) 2–3 mm long .... *A. odoratum*

***Anthoxanthum odoratum* L.**
Sweet vernalgrass
Habitat/Range: Rocky bluffs, lawns, fields and waste areas; common on SE Vancouver Island, rare N to the Queen Charlotte Islands; introduced from Eurasia.

***Anthoxanthum puelii* Lec. & Lam.** (*A. aristatum* Boiss.)
Annual vernalgrass
Habitat/Range: Fields, roadsides and waste areas; rare in SW and SC BC; introduced from Europe.

APERA

***Apera interrupta* (L.) Beauv.** (*Agrostis interrupta* L.)
Dense silky bentgrass
Habitat/Range: Dry waste areas; rare in S BC; introduced from Europe.

ARCTAGROSTIS

Arctagrostis latifolia **(R. Br.) Griseb. in Ledeb. var. *arundinacea* (Trin.) Griseb. in Ledeb.**
Polargrass
Habitat/Range: Mesic meadows and forests from the montane to subalpine zones; common in N BC north of 57°N; circumpolar, N to AK, YT and NT, E to N PQ; Greenland, Iceland, N Eurasia.

ARCTOPHILA

Arctophila fulva **(Trin.) Anderss.**
Pendantgrass
Habitat/Range: Lake and pond margins and moist streambanks in the montane to subalpine zones; rare in NW BC; circumpolar, N to AK, YT and NT, E to Labr., Greenland, N Eurasia.

ARISTIDA

1. Plants annuals; culms solid .... *A. oligantha*
1. Plants perennials; culms hollow .... *A. longiseta*

Aristida longiseta **Steud. var. *robusta* Merr.** (*A. purpurea* auct. non Nutt., *A. fasciculata* auct. non Torr.)
Red three-awn
Habitat/Range: Dry sites in the steppe and lower montane zones; common in SC BC, less frequent eastward in S BC; E to SW MB and S to IA, MN, TX and NM.

Aristida oligantha **Michx.**
Prairie three-awn
Habitat/Range: Dry disturbed sites; rare in SW BC, known only from Vancouver (probably introduced in BC); S to AZ and CA.

ARRHENATHERUM

Arrhenatherum elatius **(L.) Beauv. ex Presl & Presl**
Tall or false oatgrass
Habitat/Range: Fields and disturbed sites; common on S Vancouver Island and the Gulf Islands, rare E to SC BC; introduced from Eurasia.

AVENA

1. Lemmas usually hirsute, thin and membraneous at the tip, with geniculate awns on the lower two .... *A. fatua*
1. Lemmas glabrous, greenish and thickened at the tip, with non-geniculate awns on the lower ones, or awnless .... *A. sativa*

Avena fatua **L.**
Wild oat
Habitat/Range: Fields and waste areas; infrequent in S BC; introduced from Eurasia.

Avena sativa **L.**
Common oat
Habitat/Range: Roadsides, railways and waste areas; infrequent in S BC; introduced from Eurasia.

BECKMANNIA

***Beckmannia syzigachne* (Steud.) Fern. ssp. *baicalensis* (Kusnez.) Koyama & Kawano** (*B. erucaeformis* auct. non [L.] Host, *B. erucaeformis* [L.] Host var. *uniflora* Scribn.)

American sloughgrass

Habitat/Range: Pond margins, marshes, wet meadows and ditches in the lowland, steppe and montane zones; frequent, but scattered throughout BC east of the Coast-Cascade Mountains, absent from NW BC; N to AK, YT and NT, E to NF and S to PA, MO, NY, NM and CA.

BOUTELOUA

***Bouteloua gracilis* (HBK) Lag. ex Steud.** (*Chondrosum gracile* HBK)

Blue grama

Habitat/Range: Dry sites in the steppe zone; rare in SC and SE BC; E to S MB and S to MO, IL, TX, MX and CA.

BRIZA

***Briza maxima* L.**

Big quaking grass

Habitat/Range: Dry disturbed sites and waste areas; rare garden escape in SW BC, (known only from Horseshoe Bay); introduced from Europe.

Notes: The two similar, excluded species, *B. media* and *B. minor*, have not been collected for over 50 years. The latter have shorter spikelets - about 3 mm long in *B. minor* and up to 7 mm in *B. media*. All three species probably do not persist in our area.

BROMUS[20]

Notes: The genus *Bromus*, as with many of our complex genera, has received markedly varying taxonomic treatments. This confusion is perpetuated by a few taxonomists who give taxonomic recognition to many of the minor morphological variants within a species. In concurrence with the majority of taxonomists, formal recognition is not given to the following species: *B. aleutensis, B. canadensis,* and *B. porteri.*

1. Plants introduced annuals; lemmas bifid, the teeth over 0.5 mm long (except in *B. briziformis* which has inflated, short [1 mm long or less] or awnless lemmas).
    2. Lemmas narrowly acute to accuminate, the narrow apical teeth usually more than 2 mm long, the awns 10–60 mm long; first glume 1-nerved.
        3. First glume 4–6 (9) mm long; lemma awns 10–18 mm long .... *B. tectorum*
        3. First glume 6–20 mm long; lemma awns 18–60 mm long.
            4. First glume 6–14 mm long; awns 15–30 mm long .... *B. sterilis*
            4. First glume 13–20 (22) mm long; awns (30) 35 (60) mm long .... *B. rigidus*
    2. Lemmas rounded at the tips, usually emarginate, the teeth usually less than 1 mm long, the awns 0.5–12 (14) mm long or absent; first glume 3–5 nerved.
        5. Lemmas obtuse, inflated, 2.8–4 mm wide, awnless or awns scarcely 1 mm long .... *B. briziformis*

---

20 Key adapted from Hitchcock *et al.* (1969).

5. Lemmas somewhat dorsoventrally compressed, not inflated, 1.2–2.5 mm wide (sometimes wider in *B. squarrosus*), awns 0.5–12 (14) mm long.

6. Panicle dense or narrow, usually erect.

7. Lemmas papery, wth prominent veins, anthers less than 1 mm long .... *B. hordeaaceus*

7. Lemmas rough, without prominent veins; anthers 1.5–3 mm long ..... *B. racemosus*

6. Panicle lax, spreading or drooping.

8. Lemmas rounded, the margins usually inrolled; leaf sheaths usually glabrous .... *B. secalinus*

8. Lemmas somewhat flattened, the margins not inrolled; leaf sheaths pilose.

9. Panicle branches stiffly spreading or drooping, not flexuous; awns straight .... *B. commutatus*

9. Panicle branches lax or flexuous (often secund in *B. squarrosus)*; awns divaricate, flattened at base.

10. Spikelets somewhat laterally compressed; lemmas up to 7 mm wide; first glume 5-nerved .... *B. squarrosus*

10. Spikelets terete; lemmas up to 4 mm wide; first glume 3-nerved .... *B. japonicus*

1. Plants native perennials (except the introduced perennial *B. inermis* ssp. *inermis*); lemmas minutely bifid, teeth 0.5 mm long or less.

11. Spikelets strongly compressed; lemmas more or less keeled.

12. Panicle branches mostly spreading or drooping; auricles absent .... *B. sitchensis*

12. Panicle branches erect; auricles usually present but often minute .... *B. carinatus*

11. Spikelets not strongly compressed; lemmas rounded on the back (sometimes less rounded in *B. pacificus*).

13. Plants rhizomatous; awns 1–2 (4) mm long or awnless .... *B. inermis*

13. Plants non-rhizomatous; awns 1.5–12 mm long.

14. Lemmas pubescent evenly over the entire surface, sometimes the margins with longer hairs; glumes usually pubescent.

15. Ligules 0.5–1.5 (2) mm long; awns 1.5–3 (4) mm long; plants occurring E of the Coast-Cascade Mountains .... *B. anomalus*

15. Ligules 2–5 mm long; awns (4) 7–12 mm long; plants occurring W of the Coast-Cascade Mountains .... *B. pacificus*

14. Lemmas pubescent on the margins and usually on the lower portions of the back; glumes glabrous or with scabrous nerves or keel.

16. Ligules (2) 3–5 mm long; awns (3) 5–8 mm long .... *B. vulgaris*

16. Ligules 0.5–1 (1.5) mm long; awns 2–4 (5) mm long.

17. Culm nodes mostly glabrous; lemmas, at least the uppermost, often with sparse, short appressed hairs on the back .... *B. richardsonii*

17. Culm nodes pubescent; lemmas glabrous or scabrous on the back .... *B. ciliatus*

***Bromus anomalus* Rupr. ex Fourn.** (*B. ciliatus* var. *porteri* [Coult.] Rydb., *B. porteri* [Coult.] Nash)
Nodding brome
Habitat/Range: Mesic to dry meadows and open forests in the steppe to lower alpine zones; infrequent in NE, WC, SC and SE BC; E to MB and S to TX, NM, AZ, CA and MX.
Notes: *Bromus porteri* may better be treated as a separate species of the S US and MX.

***Bromus briziformis* Fisch. & Mey. in Fisch., Mey. & Trautv.**
Rattle brome or rattle chess
Habitat/Range: Roadsides and waste places; rare ornamental escape in SE BC, (known only from Trail and Kootenay); introduced from Eurasia.

***Bromus carinatus* Hook. & Arn.** (*B. breviaristatus* [Hook.] Thurb. in Torr., *B. carinatus* var. *linearis* Shear and var. *marginatus* [Nees] A.S. Hitchc., *B. hookerianus* Thurb. in Torr., *B. marginatus* Nees in Steud., *B. sitchensis* var. *marginatus* [Nees] Boivin)
California brome
Habitat/Range: Moist to dry slopes, meadows and open forests in all but the alpine zone; common in S BC, less frequent northward; introduced in AK and YT, E to SW SK and S to CO, NM, TX, CA and MX.
Notes: This is an extremely variable species consisting of numerous phases with mainly sympatric ranges. Two phases, var. *linearis* (with copiously pilose sheaths and blades) and var. *marginatus* (with shorter awns), are sometimes given formal recognition but there is no sound evidence that they are taxonomically different.

***Bromus ciliatus* L.** (*B. canadensis* Michx)
Fringed brome
Habitat/Range: Wet streambanks and lake margins to mesic meadows and open forests in the montane to subalpine zones; frequent in BC in and E of the Coast-Cascade Mountains; N to AK, YT and NT, E to NF and S to NJ, TN, MO, NM and CA.

***Bromus commutatus* Schrad.**
Meadow brome
Habitat/Range: Mesic to dry disturbed sites, meadows and open forests in the lowland, steppe and montane zones; infrequent in S BC; introduced from Europe.

***Bromus hordeaceus* L.** (*B. mollis* L. = ssp. *hordeaceus*, *B. thominii* Hardouin = ssp. *thominii*)
Soft brome or chess
Habitat/Range: Fields, roadsides, disturbed sites, rocky bluffs, open forests, lakeshores and dry salt marshes in the lowland, steppe and montane zones; frequent (ssp. *hordeaceus*) or infrequent (ssp. *thominii*) in SW and SC BC, rare on the Queen Charlotte Islands; introduced from Europe.
Notes: Two subspecies are recognized in BC.

1. Lemmas 6.5–7.5 mm long; glumes 4.5–6 mm and 6–6.6 mm long .... ssp. *thominii* (Hardouin) Marie & Weiller

1. Lemmas 8–11 mm long; glumes (6) 7–9 mm and (7) 8–9 mm .... ssp. *hordeaceus*

The hybrid *B. hordeaceus* ssp. *hordeaceus* x *B. lepidus* Holmberg (*B. x pseudothominii* P.M. Smith in Smith [1968]) appears to be infrequent on S Vancouver Island and rare elsewhere in S BC and is difficult to distinguish.

***Bromus inermis* Leys.** (*B. arcticus* Shear in Scribn. & Merr., *B. pumpellianus* Scribn. and *B. pumpellianus* var. *tweedyi* Scribn. in Beal = var. *pumpellianus*)

Smooth or Hungarian brome (ssp. *inermis*), Pumpelly brome (ssp. *pumpellianus*)

Habitat/Range: Moist to dry disturbed sites, fields, meadows and forests in all but the alpine zone; ssp. *inermis* - infrequent in SC and SE BC, rare northward and westward, ssp. *pumpellianus* - frequent in BC east of the Coast-Cascade Mountains; ssp. *inermis* - introduced from Eurasia; ssp. *pumpellianus* - N to AK, YT and NT, E to AB and ON and S to CO.

Notes: Two subspecies occur in BC. Numerous other variants have been recognized but all are based on minor morphological differences. The two subspecies hybridize frequently.

1. Lemmas glabrous or scabrid at the margins; awns 0.5–1 (3) mm long or absent; glumes, at least the first, tapered from the base .... ssp. *inermis*

1. Lemmas pubescent; awns (1) 2–3 (6) mm long; glumes broadened above the base .... ssp. *pumpellianus* (Scribn.) Wagnon

***Bromus japonicus* Thunb.**

Japanese brome

Habitat/Range: Fields and waste places in the steppe zone; infrequent in SC and SE BC; introduced from Eurasia.

***Bromus pacificus* Shear**

Pacific brome

Habitat/Range: Moist to mesic coastal meadows and forests in the lowland zone; infrequent in coastal BC; N to SW AK and S to OR.

***Bromus racemosus* L.**

European smooth brome

Habitat/Range: Disturbed sites and waste places; rare in S BC; introduced from Eurasia.

***Bromus richardsonii* Link**

Richardson's brome

Habitat/Range: Mesic meadows and forests in the montane and subalpine zones; common in BC in and E of the Coast-Cascade Mountains; N to AK and YT and S to ND, CO, AZ, NM and CA.

***Bromus rigidus* Roth** (*B. maximus* Desf.)

Rip-gut brome

Habitat/Range: Roadsides and waste places; infrequent on S Vancouver Island and the Gulf Islands; introduced from Europe.

***Bromus secalinus* L.**

Rye brome

Habitat/Range: Roadsides, fields and waste places; rare in WC and S BC, (known recently only from Hazelton); introduced from Eurasia.

***Bromus sitchensis* Trin.** (*B. aleutensis* Trin. ex Griseb. in Ledeb.)

Alaska brome

Habitat/Range: Moist to mesic meadows and open forests in the lowland, montane and subalpine zones; frequent in BC west of the Coast-Cascade Mountains, rare in SE BC; N to AK and S to NW OR.

Notes: The treatment here follows that of Calder and Taylor (1968) who suggested that *B. aleutensis* was merely a depauperate form of *B. sitchensis*. They are separated on minor and overlapping differences in panicle characteristics.

***Bromus squarrosus* L.**

Corn brome

Habitat/Range: Fields, disturbed sites and waste places; rare in WC, SC and SE BC; introduced from Eurasia.

***Bromus sterilis* L.**
Barren brome
Habitat/Range: Roadsides, waste places and rocky forest openings; common on S Vancouver Island and Gulf Islands, rare in SC and SE BC; introduced from Eurasia.

***Bromus tectorum* L.**
Cheatgrass
Habitat/Range: Roadsides, disturbed sites, waste places in the lowland, steppe and montane zones; common in S BC, rare elsewhere; introduced from Eurasia.

***Bromus vulgaris* (Hook.) Shear** (*B. eximius* [Shear] Piper)
Columbia brome
Habitat/Range: Moist to mesic meadows and open forests in the lowland and montane zones; common on S Vancouver Island and the Gulf Islands, rare eastward in S BC; E to AB and S to MT, WY and CA.

CALAMAGROSTIS[21]

1. Awns exserted beyond glume tips, twisted and bent.
    2. Glumes 4–5 mm long; leaf sheaths usually hairy on collars; awns exserted less than 2 mm beyond glume tips .... *C. rubescens*
    2. Glumes 4.5–8 mm long; leaf sheaths glabrous to somewhat scabrous; awns exserted 1–4 (5) mm beyond glume tips.
        3. Leaf blades smooth to scabrous above leaf sheaths, glabrous on collars; glumes glabrous to scabrous only on keel; plants of coastal BC .... *C. sesquiflora*
        3. Leaf blades long-scabrous or puberulent above leaf sheaths, glabrous to scabrous on collars; glumes scabrous; plants of interior BC .... *C. purpurascens*
1. Awns not reaching glume tips or barely exserted, straight or twisted and bent.
    4. Panicles loose, branches spreading or ascending.
        5. Glumes 5–7 (8) mm long; awns stout; callus hairs sparse, less than 1/2 length of lemmas; robust plants of coastal BC .... *C. nutkaensis*
        5. Glumes 3–5 mm long; awns slender; callus hairs copious, 1/2 to as long as lemmas; plants of interior BC, rarely coastal.
            6. Culms 3–9 dm tall, unbranched, with 2–3 nodes; panicles 4–12 cm long, with longer branches to 4 cm long; leaf blades scabrous above, smooth below .... *C. lapponica*
            6. Culms 6–15 dm tall, sometimes branched above, with 3–8 nodes; panicles 10–25 cm long, with longer branches 3–8 cm long; leaf blades scabrous above and below .... *C. canadensis*
    4. Panicles contracted, branches appressed.
        7. Callus hairs mostly 1/2 to as long as lemmas; awns straight to somewhat twisted and bent; glumes 2–4.5 mm long .... *C. stricta*
        7. Callus hairs mostly less than 1/2 the length of lemmas; awns slightly to strongly twisted and bent; glumes 4–5 mm long.
            8. Culms 1.4–4 dm tall; panicles dense, 4–8 cm long, with longer branches to 2 cm long; leaf sheaths glabrous on collar .... *C. montanensis*

---

[21] Contributed by C.W. Greene and George W. Douglas.

8. Culms 6–10 dm tall; panicles dense to somewhat open, 6–15 cm long, with longer branches 1.5–4 cm long; at least some leaf collars with tufted hairs .... *C. rubescens*

Calamagrostis canadensis **(Michx.) Beauv.** (*C. canadensis* vars. *acuminata* Vasey ex Rydb. & Shear and *robusta* Vasey in Rothr., *C. pallida* Vasey & Scribn. ex Vasey non Nutt. ex A. Gray and *C. scribneri* Beal = var. *canadensis; C. lactea* Beal and *C. scabra* J.S. Presl in C.B. Presl = var. *langsdorffii* [Link] Inman)

Bluejoint or Canada reedgrass

Habitat/Range: Moist to wet bogs, meadows and open forests in the lowland to subalpine zones; var. *canadensis* - common throughout BC, except rare in SW BC, var. *langsdorffii* - common in N BC east of the Coast-Cascade Mountains, less frequent southward; circumboreal, N to AK, YT, NT, E to NF and S to NH, NY, MI, NC, CO, NM and CA, Eurasia.

Notes: Two intergrading varieties are recognized in BC and the delimitation of each is arbitrary. The var. *langsdorfii* also intergrades with *C. lapponica*. The two varieties may be separated as follows:

1. Glumes 3–4 mm long, slightly scabrous and thin, not hispid on the keels, hyaline on the margins; awns thin, not prominent .... var. *canadensis*

1. Glumes 4–6 mm long, strongly scabrous, hispid on the keels, usually not hyaline on the margins; awns thick, more prominent .... var. *langsdorfii* (Link) Inman

Calamagrostis lapponica **(Wahlenb.) Hartm.** (*C. lapponica* var. *nearctica* Pors.)

Lapland reedgrass

Habitat/Range: Mesic to moist sites in the subalpine to alpine zones; infrequent in BC east of the Coast-Cascade Mountains; circumboreal, N to AK, YT and NT and E to NF; Greenland, Eurasia.

Calamagrostis montanensis **(Scribn.) Scribn. in Vasey**

Plains reedgrass

Habitat/Range: Dry sites in the steppe and montane zones; rare in SE and NE BC; E to MB and S to SD and WY.

Calamagrostis nutkaensis **(J.S. Presl in C.B. Presl) Steud.**

Pacific reedgrass

Habitat/Range: Moist to dry meadows, coastal bluffs, open forests and sand dunes in the lowland to subalpine zones; frequent in coastal BC; N to AK and S to CA.

Calamagrostis purpurascens **R. Br. in Richards.** (*C. purpurascens* var. *laricina* Louis-Marie)

Purple reedgrass

Habitat/Range: Mesic to dry, rocky or sandy sites in the upper montane to alpine zones; common in BC east of the Coast-Cascade Mountains; amphiberingian, N to AK, YT and NT, E to PQ and Labr. and S to CA, Greenland, E Asia.

Calamagrostis rubescens **Buckl.** (*C. luxurians* Rydb.)

Pinegrass

Habitat/Range: Dry meadows, rocky slopes and open forests in the steppe to subalpine zones; common in S BC east of the Coast-Cascade Mountains; E to AB and S to CO, NV and CA.

Calamagrostis sesquiflora **(Trin.) Tzvelev** (*C. purpurascens* R. Br. ssp. *tasuensis* Calder & Taylor)

One-and-a-half-flowered reedgrass

Habitat/Range: Mesic to moist rocky slopes in the subalpine to lower alpine zones; rare on NE Vancouver Island and the Queen Charlotte Islands; amphiberingian, N to AK and S to N WA; NE Asia.

Calamagrostis stricta **(Timm) Koel.** (*C. crassiglumis* Thurb. and *C. inexpansa* A. Gray vars. *brevior* [Vasey] Stebbins and *inexpansa* = subsp. *inexpansa; C. neglecta* auct. non [Ehrh.] Gaertn., Meyer & Scribn. = subsp. *stricta*)

Slimstem reedgrass

Habitat/Range: Mesic to wet meadows, lakeshores and open forests in the montane to alpine zones; ssp. *inexpansa* - common in BC east of the Coast-Cascade Mountains, infrequent in coastal BC, ssp. *stricta* frequent in BC east of the Coast-Cascade Mountains, rare in extreme SC and SE BC; ssp. *inexpansa* - amphiberingian, N to AK, YT and NT, E to NF and S to NH, NY, OH, IN, NE, IO, NM, AZ and CA; NE Asia, ssp. *stricta* - circumboreal, N to AK, YT and NT, E to Labr. and S to NH, MI and CO; Eurasia.

Notes: Two subspecies occur in BC. The ssp. *inexpansa* includes many apomictic forms and intergrades with the ssp. *stricta*. They may be separated as follows:

1. Glumes thick, often opaque; blades 3–6 mm wide; ligules 3–6 mm long .... ssp. *inexpansa* (A. Gray) C.W. Greene

1. Glumes thin; blades 2–4 mm wide, often involute; ligules 1–3 mm long .... ssp. *stricta*

Coastal forms of ssp. *inexpansa* with indurate glumes have been separated as *C. crassiglumis* but these seem to be semi-apomictic forms that easily fall within this morphologically variable species.

CALAMOVILFA

Calamovilfa longifolia **(Hook.) Scribn. in Hack. var. *longifolia***

Prairie sandgrass or prairie sand-reed

Habitat/Range: Dry sandy sites in the steppe and lower montane zones; rare in SE BC; E to W ON and S to MI, MO, KS, CO and N ID.

CATABROSA

Catabrosa aquatica **(L.) Beauv.**

Water hairgrass

Habitat/Range: Wet to moist meadows, streambanks, ponds and lakeshores in the montane zone; rare in SE BC, known only from Fernie; circumpolar, E to PQ and Labr. and S to NE, MI, CO, AZ and OR; Iceland, Eurasia.

Note: This species has been collected only once at Fernie in 1911.

CENCHRUS

Cenchrus longispinus **(Hack.) Fern.**

Burgrass, or field sandbur

Habitat/Range: Moist to dry sandy shores and disturbed sites in the steppe zone; rare in extreme SC BC; E to ON and S through most of US.

CINNA

Cinna latifolia **(Trevir. ex Gopp.) Griseb. in Ledeb.**

Nodding wood-reed

Habitat/Range: Moist streambanks, swamps, meadows and forests from the lowland to subalpine zones; common throughout BC; circumpolar, N to AK, YT and SW NT, E to NF and S to NC, TN, NM, NV and CA; Eurasia.

COLEANTHUS

Coleanthus subtilis **(Tratt.) Seidl in Roem. & Schult.**
Moss grass
Habitat/Range: Damp, muddy lake margins in the lower montane zone; rare in SC BC, (known only from Shuswap Lake); native to N and C Europe.
Notes: There is a good possibility that this species is a circumpolar disjunct and not merely an introduction since many companion plants occurring at Shuswap Lake are the same as those found in Europe (A. Ceska, pers. comm.). The small stature (2–8 cm) of the plant and its late flowering date (October) may be the reason it is known only from the above location and the lower Columbia River, OR/WA.

CYNODON

Cynodon dactylon **(L.) Pers.**
Bermuda grass
Habitat/Range: Waste places in the lowland zone; rare in the Vancouver area; introduced from the E and S United States.

CYNOSURUS

1. Panicles spike-like, less than 1 cm wide; awns less than 1 mm long; plants perennials, 4–8 dm tall; ligules 1–1.5 mm long, finely erose-ciliolate .... *C. cristatus*
1. Panicles ovoid, more than 1 cm wide; awns 3–10 mm long; plants annuals, (1) 2–5 dm tall; ligules 2–7 (11) mm long, erose to toothed .... *C. echinatus*

Cynosurus cristatus **L.**
Crested dogtail
Habitat/Range: Fields and waste areas; frequent on S Vancouver Island, less frequent E in S BC; introduced from Eurasia.

Cynosurus echinatus **L.**
Hedgehog dogtail
Habitat/Range: Sandy shores, rocky bluffs, open forests and disturbed sites; locally frequent on S Vancouver Island and the Gulf Islands, rare in SE BC; introduced from Europe.

DACTYLIS

Dactylis glomerata **L.**
Orchardgrass
Habitat/Range: Fields, meadows, roadsides and disturbed sites; common in S BC, less frequent in N BC; introduced from Eurasia.

DANTHONIA[22]

1. Lemmas less than 6 mm long .... *D. spicata*
1. Lemmas more than 6 mm long.

---

22 Contributed by S.J. Darbyshire.

2. Inflorescence branches spreading, lower pedicels as long as or longer than spikelets .... *D. californica*

2. Inflorescence branches ascending, lower pedicels shorter than to as long as spikelets.

    3. Spikelets 5–11 (15), lower panicle branch with more than one spikelet .... *D. intermedia*

    3. Spikelets 1 (3), lower panicle branch never with more than one spikelet .... *D. unispicata*

***Danthonia californica* Boland.** (*D. americana* Scribn., *D. macounii* A.S. Hitchc.)
California oatgrass
Habitat/Range: Moist to dry meadows, rocky bluffs and open forests in the lowland and montane zones; common on S Vancouver Island and the Gulf Islands, rare on the Queen Charlotte Islands, N Vancouver Island and the lower mainland; E to SC SK and S to CO and CA; Chile.

***Danthonia intermedia* Vasey** (*D. canadensis* Baum & Findlay in Findlay & Baum, *D. cusickii* [T.A. Williams] A.S. Hitchc.)
Timber oatgrass
Habitat/Range: Moist bogs and meadows to dry, rocky slopes in the lowland to alpine zones; common in BC south of 55°N, infrequent northward; N to S AK and S YT, E to SW MB, PQ and NF and S to NM and CA.

***Danthonia spicata* (L.) Beauv. ex Roem. & Schult.** (*D. canadensis* Baum & Findlay in Findlay & Baum *pro parte sed non typus*, *D. pinetorum* [Piper] Piper in Piper & Beattie, *D. sericea* Nutt. *sensu* Findlay & Baum in Findlay & Baum, *D. thermalis* Scribn.)
Poverty oatgrass
Habitat/Range: Mesic, open forests on sterile soils in the lowland and montane zones; common throughout all but NW BC; N to SE AK, S YT, E to NF and S to FL, AL, TX, NM, NE MX and N CA; Greenland.

***Danthonia unispicata* (Munro ex Thurb. in A. Gray) Munro ex Macoun**
One-spike oatgrass
Habitat/Range: Mesic to dry sites from the lowland to subalpine zones; infrequent in S BC; E to SW SK and S to CO and CA.

DESCHAMPSIA

1. Plants annuals; plants simple to somewhat tufted .... *D. danthonioides*

1. Plants perennials; plants densely tufted.

    2. Leaves flat, 1.5–3 (4) mm wide .... *D. cespitosa*

    2. Leaves filiform, 1.5 mm wide or less.

        3. Panicles narrow, with erect branches; lemma awns straight to slightly geniculate .... *D. elongata*

        3. Panicles open, with spreading branches; lemma awns twisted, geniculate .... *D. flexuosa*

Deschampsia cespitosa **(L.) Beauv.** (*Deschampsia beringensis* Hult. and *D. cespitosa* var. *arctica* Vasey = ssp. *beringensis*; *D. holciformis* J.S. Presl in C.B. Presl, *D. cespitosa* var. *littoralis* [Reut.] Richter and *D. cespitosa* var. *longiflora* Beal = ssp. *cespitosa*)

Tufted hairgrass

Habitat/Range: Saline meadows and tidal marshes in the lowland zone and moist meadows in the montane to alpine zones (ssp. *beringensis)* and moist to mesic meadows, rocky ridges and talus slopes in the montane to alpine zones (ssp. *cespitosa*); ssp. *beringensis* - common along the coast, ssp. *cespitosa* - common in and E of the Coast-Cascade Mountains throughout BC; ssp. *beringensis* - amphiberingian, N to AK and S to CA; Asia; ssp. *cespitosa* - circumpojar, N to AK, YT and NT, E to NF and S to GA, VA, IL, MN, NM, MX and CA; Eurasia.

Notes: This is a wide-spread, difficult complex and until an adequate monographic study is undertaken we are recognizing only two subspecies for BC.

1. Glumes mostly 5–7 mm long; lemmas averaging 4 mm long; panicles open during anthesis .... ssp. *beringensis* (Hult.) Lawrence

1. Glumes mostly less than 5 mm long; lemmas averaging 3.5 mm, if longer then the panicles closed at anthesis .... ssp. *cespitosa*

Deschampsia danthonioides **(Trin.) Munro ex Benth.**

Annual hairgrass

Habitat/Range: Roadsides, dry slopes and vernal pools in the lowland and montane zones; infrequent in S BC; N to AK and YT, E to SW AB and S to MT, UT, AZ, MX and CA; Chile.

Deschampsia elongata **(Hook.) Munro ex. Benth.** (*D. ciliata* [Vasey ex Beal] Rydb.)

Slender hairgrass

Habitat/Range: Moist to mesic slopes, streambanks, open forests, meadows and disturbed sites from the lowland to alpine zones; frequent in S BC; N to AK and S YT, E to SW AB and S to MT, NM, AZ, MX and CA.

Deschampsia flexuosa **(L.) Trin.**

Wavy hairgrass

Habitat/Range: Peaty soil along roadsides in the lowland zone; rare in SW BC, known only from the Vancouver area; introduced from E North America.

DIGITARIA

1. Sheaths and collars more or less strongly pilose; upper glumes as long as the pale greenish-brown fertile lemmas or less .... *D. sanguinalis*

1. Sheaths and collars glabrous or nearly so; upper glumes about as long as the dark-brownish or dark-purplish fertile lemmas .... *D. ischaemum*

Digitaria ischaemum **(Schreb. ex Schweig.) Schreb. ex Muhl.**

Smooth crabgrass

Habitat/Range: Fields, roadsides and waste areas; infrequent in SW BC (lower Fraser Valley), rare on S Vancouver Island; introduced from Eurasia.

Digitaria sanguinalis **(L.) Scop.**

Hairy crabgrass

Habitat/Range: Lawns, fields, roadsides and waste areas; infrequent in SW and SC BC; introduced from Eurasia.

DISTICHLIS

***Distichlis spicata* (L.) Greene** (*D. spicata* var. *borealis* [J.S. Presl in C.B. Presl] Beetle and var. *stolonifera* Beetle = var. *spicata*, *D. stricta* [Torr.] Rydb. and *D. dentata* Rydb. = var. *stricta*)

Alkali saltgrass (var. *stricta*) and seashore saltgrass (var. *spicata*)

Habitat/Range: Sandy lake shores and moist alkaline sites in the steppe zone (var. *stricta*) or coastal beaches and salt marshes (var. *spicata*); var. *spicata* - common in SW BC, var. *stricta* - frequent in SC BC; var. *stricta* - E to SK and S to MO, OK, TX and CA; var. *spicata* - S to CA, disjunct to E coast of North America.

Notes: Two varieties occur in BC.

1. Long hairs usually present on the collars as well as at the summits of the sheath margins; plants of the steppe zone .... var. *stricta* (Torr.) Beetle
1. Hairs usually lacking on the collars, present only on the summits of the sheath margins; plants maritime .... var. *spicata*

ECHINOCHLOA

***Echinochloa crusgalli* (L.) Beauv.**

Large barnyard-grass

Habitat/Range: Moist ditches and disturbed sites; common in S BC; introduced from Eurasia.

X ELYMORDEUM [23]

***x Elymordeum macounii* (Vasey) Barkw.** (*x Agrohordeum macounii* [Vasey] Lepage, *Elymus macounii* Vasey)

Macoun's wildrye

Habitat/Range: Dry sites in the steppe and montane zones; rare in EC, SC and SE BC, N to AK, YT and NT, E to MB and S to MN, KS, UT, NM, AZ and CA.

Notes: This taxon is a sterile hybrid between *Elymus trachycaulus* and *Hordeum jubatum*.

ELYMUS [24,25]

1. Plants with 1 spikelet at most or all nodes.
    2. Glumes with awns at least as long as the bodies of the glumes .... *E. x saundersii*
    2. Glumes awnless or with an awn shorter than the body of the glume.
        3. Plants caespitose.
            4. Anthers 1–2.5 mm long; spikelets in the middle of the inflorescence overlapping the spikelets immediately above.
                5. Glumes widest near the tip with hyaline margins above 0.5 mm wide on the sides; glume veins smooth or scabridulous .... *E. alaskanus*
                5. Glumes widest at midlength with hyaline margins about 0.3 mm wide on the sides; glume veins scabrous .... *E. trachycaulus*

---

23 Contributed by M.E. Barkworth.

24 Contributed by M.E. Barkworth.

25 Considerable controversy exists concerning the most appropriate generic treatment for the Triticeae. At the request of the editors the generic treatment adopted here is that which will be used by the forthcoming Flora North America project.

4. Anthers 4–6 mm long; spikelets in the middle of the inflorescence overlapping the next spikelets on the opposite side of the rachis but not the spikelets immediately above.

   6. Glumes thin, flexible, acute; culms 1–10 dm tall; native species found in a variety of habitats in the arid interior .... *E. spicatus*

   6. Glumes thick, stiff, obtuse to truncate; culms 7–20 dm tall; introduced, found in disturbed sites such as roadsides .... *E. elongatus*

3. Plants rhizomatous.

   7. Glumes curved slightly to one side (examine the midvein) and acuminate, tapering from midlength or below .... *E. smithii*

   7. Glumes straight, acute to obtuse, narrowing from midlength or above.

      8. Glumes shortly awned, their midveins more prominent and scabrous distally than proximally .... *E. repens*

      8. Glumes not awned, their midveins usually equally prominent and scabrous throughout or more prominent basally.

         9. Glumes and lemmas thick, stiff, obtuse; plants introduced .... *E. hispidus*

         9. Glumes and lemmas thin, flexible, acute; plants native.

            10. Glumes pubescent; paleas with at least a few hairs on the keels and some hairs between the keels (hard to see); plants of extreme NW BC .... *E. yukonensis*

            10. Glumes glabrous; paleas scabrous but not hairy on and between the keels; plants of S and NE BC .... *E. lanceolatus*

1. Plants with 2 or more spikelets at most, if not all, nodes.

   11. Rachis disarticulating at maturity; glumes 2.5–10 mm long, awnlike .... *E. elymoides*

   11. Rachis not disarticulating at maturity; glumes less than 2.5 mm long, flat or awnlike.

      12. Lemmas strongly scabrid-pubescent, tapering into a slender, divergent awn up to 35 mm long .... *E. canadensis*

      12. Lemmas glabrous or scabrous over most of their backs; unawned or with straight awn.

         13. Lemmas with marginal hairs distinctly longer than any other lemma pubescence; spikes flexous .... *E. hirsutus*

         13. Lemmas with marginal hairs little, if any, longer than other lemma hairs; spikes generally erect .... *E. glaucus*

***Elymus alaskanus* (Scribn. & Merr.) A. Löve. ssp. *latiglumis* (Scribn. & J.G. Smith) A. Löve** (*Agropyron alaskanum* Scribn. & Merr., *A. boreale* [Turcz.] Drobov ex Polunin ssp. *hyperarctium* [Polunin] Melderis, *A. caninum* var. *latiglume* [Scribn. & J.G. Smith] Pease & Moore, *A. latiglume* [Scribn. & J.G. Smith] Rydb.)

Alaskan wildrye

Habitat/Range: Mesic to dry sites in all vegetation zones; common throughout BC; N to AK, YT and NT, E to NF and S to WV, KS, NM, AZ, MX and CA.

Notes: It is possible that with further study this taxon may be treated as part of the *E. trachycaulus* complex. On the other hand, one or more taxa, now treated under *E. trachycaulus,* may prove to be part of *E. alaskanus.* At the present time there is no combination available for the treatment of the ssp. *latiglume* under *E. trachycaulus.*

***Elymus canadensis* L.** (*E. brachystachys* Scribn. & Ball)
Canada wildrye
Habitat/Range: Moist to dry sandy or gravelly meadows in the steppe and montane zones; infrequent in SC and SE BC; N to S AK and NT, E to NB and NS and S to NC, MO, TX, NM, AZ and CA.

***Elymus elongatus* (Host) Runemark ssp. *ponticus* (Podp.) Meld.** (*Agropyron elongatum* auct. non [Host] Beauv., *Elytrigia pontica* [Podp.] Holub, *Thinopyrum ponticum* [Podp.] Barkw. ssp. *ponticum*)
Tall wheatgrass
Habitat/Range: Dry roadsides, slopes and open forests in the steppe and lower montane zones; infrequent in SC and SE BC; introduced from Eurasia.

***Elymus elymoides* (Raf.) Swezey** (*Sitanion hystrix* [Nutt.] J.G. Smith)
Squirreltail grass, or squirreltail
Habitat/Range: Dry rocky slopes, roadsides and scree slopes in the steppe to alpine zones; common in S BC in and E of the Coast-Cascade Mountains; E to SK and S to TX, NM, AZ, MX and CA.
Notes: Two subspecies occur in BC.

1. All glumes entire, their awns usually shorter than the lemma awns .... ssp. *californicus* Barkw.
1. At least one glume at most nodes bifid, their awns longer than the lemma awns .... ssp. *elymoides*

***Elymus glaucus* Buckl.** (*E. glaucus* var. *breviaristatus* Davy in Jeps. and *E. virescens* Piper = ssp. *virescens*)
Blue wildrye
Habitat/Range: Moist to dry slopes, meadows and open forests in the lowland and montane to alpine zones; ssp. *glaucus* - common in S BC, less frequent N of 53°N, ssp. *virescens* - infrequent in coastal BC, rare E of the Coast-Cascade Mountains; N to SE AK, E to ON and S to AR, MI, NY, CO, NM and CA.
Notes: Two subspecies occur in BC.

1. Lemmas awnless or with awns less than 5 mm long .... ssp. *virescens* (Piper) A. Löve
1. Lemma awns 10–30 mm long .... ssp. *glaucus*

***Elymus hirsutus* J.S. Presl in C.B. Presl**
Hairy wildrye
Habitat/Range: Moist meadows, streamsides and open forests in all vegetation zones; frequent in SW and SC BC, rare in SE BC; N to AK and S to NW OR.
Notes: This species appears to hybridize with *E. glaucus*.

***Elymus hispidus* (Opiz) Meld.** (*Agropyron intermedium* [Host] Beauv., *A. trichophorum* Richt., *Elytrigia intermedia* [Host] Nevski, *Thinopyrum intermedium* [Host] Barkw. & D.R. Dewey)
Intermediate wheatgrass
Habitat/Range: Dry roadsides and forests in the montane zone; rare in SC and SE BC; introduced from Eurasia.

***Elymus lanceolatus* (Scribn. & J.G. Smith) Gould** (*Agropyron albicans* Scribn. & J.G. Smith = ssp. *albicans; A. dasystachyum* [Hook.] Scribn., *A. riparium* Scribn. & J.G. Smith = ssp. *lanceolatus*)

Thickspike wildrye

Habitat/Range: Dry alkaline flats, sand dunes, gravelly sites and open forests in the steppe and montane zones; common in NE, SC and SE BC; E to AB and S to IL, MI, CO and CA.

Notes: *Agropyron riparium* was distinguished from *A. dasystachyum* by its glabrous, or nearly glabrous lemmas, but the variation in this character is continuous. A hybrid between *E. lanceolatus* and *E. trachycaulus, E.* x *pseudorepens* (Scribn. & J.G. Smith) Barkw. & D.R. Dewey (*Agropyron x pseudorepens* Scribn. & J.G. Smith), is rare in BC. It resembles *Elymus repens* but has glumes nearly as long as the spikekets and the rachilla is pubescent. Two subspecies occur in BC.

1. Lemmas awnless or with straight awns less than 5 mm long .... ssp. *lanceolatus*

1. Lemmas awned, the awns more than 5 mm long and usually divergent .... ssp. *albicans* (Scribn. & J.G. Smith) Barkw.

***Elymus repens* (L.) Nevski** (*Agropyron repens* [L.] Beauv., *Elytrigia repens* [L.] Nevski)

Quackgrass

Habitat/Range: Roadsides, fields, gardens and disturbed sites in the lowland, steppe and montane zones; common in BC south of 52°N, less frequent N to 53°N; introduced from Eurasia.

***Elymus x saundersii* Vasey** (*Agropyron saundersii* [Vasey] A.S. Hitchc.)

Saunder's wildrye

Habitat/Range: Moist to dry sites in the montane zone; rare in NE and SC BC, S to CO and OR.

Notes: This taxon is a hybrid between *Elymus elymoides* and *E. trachycaulus.*

***Elymus smithii* (Rydb.) Gould** (*Agropyron smithii* Rydb., *Elytrigia smithii* [Rydb.] Nevski, *Pascopyrum smithii* [Rydb.] A. Löve)

Western or bluestem wheatgrass

Habitat/Range: Moist to mesic slopes in the steppe and montane zones; infrequent in BC east of the Coast-Cascade Mountains, except locally common in Peace River grasslands; E to ON and S to NY, TN, TX, NM, AZ, NV and OR.

***Elymus spicata* (Pursh) Gould** (*Agropyron spicatum* [Pursh] Scribn. & J.G. Smith, *A. spicatum* var. *inerme* [Scribn. & J.G. Smith] Heller, *Pseudoroegneria spicata* [Pursh] A. Löve)

Bluebunch wheatgrass

Habitat/Range: Dry, open sites in the steppe and montane zones; common in SC and SE BC, rare elsewhere in BC; N to AK and YT, E to SK and S to ND, TX, NM, AZ and NE CA.

Notes: Awned and unawned specimens differ by a single gene, a difference not meriting separate taxonomic recognition. Most BC plants are unawned.

***Elymus trachycaulus* (Link) Gould in Shinners** (*Agropyron x brevifolium* Scribn., *A. caninum* [L.] Beauv. ssp. *majus* [Vasey] C.L. Hitchc. vars. *andinum* [Scribn. & J.G. Smith] Pease & Moore and *unilaterale* [Vasey] C.L. Hitchc., *A. pauciflorum* vars. *glaucum* [Pease & Moore] Tayl. & MacBryde and *unilaterale* [Vasey] Tayl. & MacBryde, *A. subsecundum* [Link] A.S. Hitchc. and *Elymus pauciflorus* [Schwein. in Keat] Gould var. *subsecundus* Gould = ssp. *subsecundus*; *Agropyron novae-angliae* Scribn. ex Braun *et al., A. pauciflorum* [Schwein. in Keat] A.S. Hitchc. in Silveus var. *novae-angliae* [Scribn. ex Braun *et al.*] Tayl. & MacBryde, *A. caninum* auct. non [L.] Beauv., *A. caninum* ssp. *majus* var. *majus, A. pauciflorum, A. scribneri* Vasey, *A. trachycaulum* Link, *A. violaceum* [Hornem.] Lange in Reinhardt *et al., Elymus scribneri* [Vasey] M.E. Jones, and *E. trachycaulus* ssp. *violaceus* [Hornem.] A. Löve & D. Löve = ssp. *trachycaulus)*

Slender wheatgrass

Habitat/Range: Mesic to dry sites in all vegetation zones; spp. *subsecundus* - frequent in S BC, less frequent northward; ssp. *trachycaulus* - common throughout BC; N to AK, YT and NT, E to NF and S to WV, KS, NM, AZ, MX and CA.

Notes: The *Elymus trachycaulus* complex is a taxonomic nightmare. C.L. Hitchcock treated *Elymus trachycaulus* as part of a circumboreal species, *Agropyron caninum,* and recognized five varieties in the Pacific Northwest. Other taxonomists have recognized more varieties (or subspecies), others fewer. It appears to form at least partially fertile hybrids with a number of other species, including both *E. alaskanus* and *E. lanceolatus*, and some of these hybrids have been named as species. The treatment adopted here is conservative in the number of species and subspecies accepted.

1. Lemmas awned, the awns 10–30 mm long .... ssp. *subsecundus* (Link) Gould
1. Lemmas unawned or the awns less than 10 mm long .... ssp. *trachycaulus*

***Elymus yukonensis* (Scribn. & Merr.) A. Löve & D. Löve** (*Agropyron yukonensis* auct. non Scribn. & Merr., *Elymus lanceolatus* ssp. *yukonensis* [Scribn. & Merr.] A. Löve)
Yukon wildrye
Habitat/Range: Mesic to dry, sometimes alkaline, sandy or gravelly river terraces and floodplains in the montane zone; rare in NW BC; N to AK and YT.

ERAGROSTIS[26]

1. Craterlike glands on leaf blade margins, spikelet pedicels, and keels of lemmas and glumes.
   2. Spikelets 2–4 mm wide at maturity, 10–40 flowered, commonly more than 11 mm long; lemmas 1.8–2.8 mm long, bearing 2–5 conspicuous glands along the midnerve .... *E. cilianensis*
   2. Spikelets 1.5–2.2 mm wide, 7–12 (20) flowered, 4–11 mm long; lemmas 1.3–2.0 (2.2) mm long, occasionally bearing 1 or 2 inconspicuous glands along the midnerve .... *E. minor*
1. Craterlike glands not present on leaf blade margins, spikelet pedicels, or keels of lemmas and glumes.
   3. First glume more than half as long as the lowest lemma, 0.5–1.1 mm long; paleas very persistent; panicle branches solitary or paired on lowest 2 nodes; lemmas with conspicuous lateral nerves .... *E. pectinacea*
   3. First glume less than half as long as the lowest lemma, 0.3–0.8 mm long; paleas easily deciduous; panicle branches commonly whorled on the lowest 2 nodes; lemmas with inconspicuous lateral nerves .... *E. pilosa*

***Eragrostis cilianensis* (All.) Mosher** (*E. megastachya* [Koel.] Link)
Stinkgrass
Habitat/Range: Moist to dry roadsides and sandy waste places in the steppe zone; infrequent in S BC east of the Coast-Cascade Mountains; introduced from Eurasia.

***Eragrostis minor* Host** (*E. poaeoides* Beauv. ex Roem. & Schult.)
Little lovegrass
Habitat/Range: Moist to dry roadsides and waste places in the steppe zone; infrequent in S BC east of the Coast-Cascade Mountains; introduced from Eurasia.

***Eragrostis pectinacea* (Michx.) Nees**
Tufted lovegrass
Habitat/Range: Moist to dry roadsides and waste places in the steppe zone; rare in SC BC, known only from Osoyoos; disjunctly E to ON, PQ and NS and S to FL, TX, MX, AZ and NM.
Notes: *Eragrostis pectinacea* is often difficult to separate from *E. pilosa*. The lemmas of *E. pectinacea* are usually longer (1.0–2.2 mm) than those of *E. pilosa* (1.2–1.8 mm).

---

26 Contributed by P.M. Peterson.

Eragrostis pilosa **(L.) Beauv.**
India lovegrass
Habitat/Range: Wet to moist roadsides and waste places in the steppe zone; infrequent in S BC east of the Coast-Cascade Mountains; introduced from Europe.

FESTUCA[27]

1. Leaf sheaths with auricles.
   2. Auricles with hairs; lemma awns 0.1–3.0 mm long .... *F. arundinacea*
   2. Auricles without hairs; lemma awns 0.0–0.2 mm long .... *F. pratensis*
1. Leaf sheaths without auricles.
   3. Rhizomes usually present; leaf sheaths reddish, often splitting into fibres, often with retrorse hairs .... *F. rubra*
   3. Rhizomes usually absent; leaf sheaths usually not shredding into fibres.
      4. Lemma awns usually more than 2.5 mm long.
         5. Leaf blades flat or loosely rolled, (2) 3–10 mm wide.
            6. Lemma callus elongated with a tuft of stiff hairs at the base; ligules ciliate with hairs usually 0.5–1.0 x the length of the basal membranous part .... *F. subuliflora*
            6. Lemmas callus not elongated, without hairs at the base; ligules ciliolate with hairs much shorter than 0.5 x the length of the basal membranous part .... *F. subulata*
         5. Leaf blades folded to appear less than 2 mm wide.
            7. Rachilla usually somewhat zig-zag and visible between the florets; anthers 2.8–4.5 mm long; ovary tops glabrous .... *F. idahoensis*
            7. Rachilla straighter, usually not visible between the florets; anthers 1.4–3 mm long; ovary tops hairy .... *F. occidentalis*
      4. Lemma awns usually less than 2.5 mm long.
         8. Culm nodes not exposed; bunchgrasses more than 25 cm high.
            9. Lower panicle branches weak, to 10 cm long, often reflexed and often secund; spikelets usually strongly purplish; lemma margins hyaline, giving a sheen to the spikelets .... *F. altaica*
            9. Lower panicle branches often rigid, 2–7 cm long, erect or ascending, not secund; spikelets usually pale green below and subdued purplish above; lemma margins not conspicuously hyaline .... *F. campestris*
         8. Culm nodes exposed and/or plants less than 25 cm tall.
            10. Spikelets, or most of them, vegetatively proliferating .... *F. viviparoidea*
            10. Spikelets not vegetatively proliferating.
                11. Anthers 2 mm long or more.
                    12. Anthers 3–4 mm long; ovary top hairy; infrequent native in subalpine and alpine meadows of SC and SE BC .... *F. viridula*

---

27 Contributed by S.G. Aiken and L.E. Pavlick.

12. Anthers 2–3 mm long; ovary top glabrous; rare introduction in SE BC .... *F. trachyphylla*

11. Anthers less than 2 mm long.

13. Lemmas finely pointed, rarely awned .... *F. filiformis*

13. Lemmas awned.

14. Leaf blades more or less pyriform or terete in cross-section, with strongly developed sclerenchyma strands .... *F. saximontana*

14. Leaf blades angular and more or less irregular-hexagonal in cross-section, with weakly developed sclerenchyma strands.

15. Culms densely pubescent near the inflorescence .... *F. baffinensis*

15. Culms mostly glabrous.

16. Lowermost lemmas 3.0–4.2 mm long; leaf blades with a folded width of 0.35–0.6 mm; ovary tops glabrous .... *F. brachyphylla*

16. Lowermost lemmas 2.8–3.3 mm long; leaf blades with a folded width of 0.2–0.4 mm; ovary tops with a few stiff hairs .... *F. minutiflora*

***Festuca altaica* Trin. in Ledeb.** (*F. scabrella* Torr. in Hook.)
Altai fescue or northern rough fescue
Habitat/Range: Mesic to dry meadows and forest openings from the montane (N BC) to alpine zones; common in BC east of the Coast Mountains crest, S to 51°N; amphiberingian; N to AK, YT and NT, disjunct in N MI, PQ and NF; E Asia.

***Festuca arundinacea* Schreb.** (*F. elatior* L., *F. elatior* var. *arundinacea* [Schreb.] Wimm.)
Tall fescue
Habitat/Range: Beaches, roadsides, disturbed areas and moist meadows; infrequent in SW BC, rare eastward in S BC; introduced from Europe.
Notes: May be more widespread than the limited number of herbarium collections from S BC suggest. It is now widely planted in other parts of North America for stabilizing roadsides and as a pasture grass.

***Festuca baffinensis* Polun.**
Baffin fescue
Habitat/Range: Mesic to dry rocky areas in the high alpine zone; frequent in NE and E BC, rare in the Coast Mountains of N and WC BC; circumpolar, N to AK, YT and NT, E to W AB and S to CO, disjunct in PQ and E NT; Greenland, Eurasia.

***Festuca brachyphylla* Schult. and Schult. fil.** (*F. ovina* L. var. *brevifolia* [R. Br.] S. Wats. and var. *brachyphylla* [Schult. & Schult. fil.] Piper)
Alpine fescue
Habitat/Range: Mesic to dry rocky slopes, scree slopes and moraines in the alpine zone; frequent throughout BC in and E of the Coast-Cascade Mountains; circumpolar, N to AK, YT and NT, E to NF and S to CO, Greenland, Eurasia.

***Festuca campestris* Rydb.** (*F. scabrella* Torr. in Hook. var. *major* Vasey, *F. altaica* ssp. *scabrella* [Torr. in Hook.] Hult.)
Rough fescue
Habitat/Range: Mesic to dry meadows and forest openings in the steppe to subalpine zones; common in SC and SE BC south of 51°N; E to SW AB and S to ND, CO, ID and OR.

***Festuca filiformis* Pourr.** (*F. capillata* Lam., *F. tenuifolia* Sibth., *F. ovina* L. var. *capillata* [Lam.] Alefield)
Hair fescue
Habitat/Range: Dry roadsides, disturbed sites and lawns; rare, known only from Terrace and Vancouver; introduced from Europe.
Notes: Sold as "Hair fescue" and *F. ovina*, it is perhaps more widespread in lawns but has rarely been collected.

***Festuce idahoensis* Elmer** (*F. idahoensis* var. *roemeri* Pavl., *F. roemeri* [Pavl.] Alexeev)
Idaho fescue
Habitat/Range: Mesic grasslands from the lowland to subalpine zones; frequent in BC E of the Coast-Cascade Mountains and S of 50°N, and on S Vancouver Island and the Gulf Islands; E to S SK and S to CO and CA.
Notes: This is an extremely variable taxon which is sometimes treated as two separate species; plants from west of the Cascade Mountains have been distinguished as *F. roemeri* (Pavl.) Alexeev.

***Festuca minutiflora* Rydb.**
Little fescue
Habitat/Range: Dry, stony slopes in the alpine zone; rare in the Coast-Cascade Mountains; N to YT and S to AZ and CA.
Notes: Only recently recognized from BC, it is known from at least 6 sites and is possibly more common. In the past it has been mistaken for *F. brachyphylla*.

***Festuca occidentalis* Hook.**
Western fescue
Habitat/Range: Dry to mesic forests and forest openings in the lowland and montane zones; frequent in S BC, infrequent northward to 57°N, and on the Queen Charlotte Islands; E to SW AB and S to WY, UT and CA, disjunct in ON and MI.

***Festuca pratensis* Hudson**
Meadow fescue
Habitat/Range: Moist pastures, roadsides, and lakesides; infrequent in SC BC, rare in SW BC and EC BC (Prince George); introduced from Europe.
Notes: Perhaps more widespread, but infrequently collected. This was a popular grass for planting earlier this century in other places in North America, but it is now being replaced by tall fescue in similar situations.

***Festuca rubra* L. sens. lat.** (*F. prolifera* auct. non Piper, *F. rubra* var. *richardsonii* [Hook.] Hult., *F. richardsonii* Hook., *F. rubra* ssp. *arenicola* Alexeev, *F. rubra* ssp. *aucta* [Krecz. & Bobr.] Hult., *F. rubra* ssp. *aucta* f. *pseudovivipara* Pavl., *F. rubra* var. *littoralis* Vasey ex Beal, *F. rubra* ssp. *pruinosa* [Hack.] Piper, *F. rubra* ssp. *secunda* [J.S. Presl in C.B. Presl] Pavl., *F. rubra* ssp. *secunda* var. *mediana* Pavl., *F. rubra* ssp. *vallicola* [Rydb.] Pavl.)
Red fescue
Habitat/Range: Wet to mesic beaches, streambanks, meadows, gravelly sites, disturbed areas, roadsides and fields in the lowland to alpine zones; common throughout BC; circumpolar, N to AK, YT, NT, E to NF, and S to TN, NC, TX, NM and CA; Greenland, Iceland, Eurasia.
Notes: This polymorphic species complex is treated here as a single species for convenience. Some distinctive native and introduced forms have been recognized as separate species or treated at infraspecific rank; however, transitional forms occur between some of them (Pavlick 1985). Over 100 cultivated varieties of this complex are sold in Europe; some have been introduced into North America where hybridization with native forms further obscures the taxonomic picture.

Festuca saximontana **Rydb.** (*F. brachyphylla* var. *rydbergii* [St. Yves] Cronq. *F. ovina* L. var. *saximontana* [Rydb.] Gleas., *F. saximontana* var. *purpusiana* [St.-Yves] Freder. & Pavl., *F. saximontana* var. *robertsiana* Pavl.)
Rocky Mountain fescue
Habitat/Range: Mesic to dry meadows and forest openings in the montane to lower alpine zones; common in BC in and E of the Coast-Cascade Mountains, infrequent in SW BC; N to AK, YT, NT, E to NF and S to WI and CA.
Notes: *F. saximontana* var. *purpusiana* is the name given to subalpine and lower alpine forms which are 8–20 (25) cm high, in which the culms are about 2–3 times the length of the basal tufts and the spikelets are moderately purplish.

Festuca subulata **Trin. in Bong.**
Bearded fescue
Habitat/Range: Moist to mesic forests and forest openings in the lowland and montane zones; frequent in SW BC and on the Queen Charlotte Islands, infrequent eastward in S BC; N to SW AK, E to W AB and S to UT, NV and CA.

Festuca subuliflora **Scribn. in Macoun**
Crinkle-awned fescue
Habitat/Range: Mesic forests; frequent in the lowland zone of Vancouver Island, Gulf Islands, and the lower mainland; S along the coast to CA.

Festuca trachyphylla **(Hack.) Kraj.** (*F. brevipila* Tracey, *F. duriuscula* auct. non L., *F. ovina* L. var. *duriuscula* auct. non L.)
Hard fescue
Habitat/Range: Disturbed areas; rare in S BC; introduced from Europe.
Notes: Several cultivars of this species are grown in Holland and distributed widely.

Festuca viridula **Vasey**
Green fescue
Habitat/Range: Mesic to dry slopes in the upper subalpine and alpine zones; infrequent in SC and SE BC south of 50°N; E to AB and S to MT and CA.

Festuca viviparoidea **Kraj. ex Pavl. ssp.** ***krajinae*** **Pavl.** (*F. vivipara* [L.] Sm. ssp. *glabra* Freder.)
Viviparous fescue
Habitat/Range: Mesic to dry rocky slopes and outcrops in the alpine zone; infrequent in SE and NE BC, rare in NW BC; N to AK, YT and NT, E to SW AB.
Notes: It is not known whether the pseudovivipary of this taxon is environmentally induced or genetically controlled; it may be a hybrid between *F. baffinensis* and *F. brachyphylla* or even *F. rubra*.

GLYCERIA[28]

1. Spikelets 10–15 mm long, linear in outline.
    2. Lowest lemmas 5–6 mm long; anthers 1.1–1.6 mm long .... *G. occidentalis*
    2. Lowest lemmas 3–4 mm long; anthers 0.5–1.0 mm long.
        3. Blades papillose on the adaxial surface .... *G. borealis*
        3. Blades scabridulous on the adaxial surface .... *G. leptostachya*
1. Spikelets 2.5–6.5 mm long, ovate to oblong in outline.

---

28 Contributed by M.E. Barkworth.

4. Nerves of the lemmas not prominently raised; spikelets 3–5 mm wide; paleas often exposed .... *G. canadensis*
4. Nerves of the lemmas prominently raised; spikelets at most about 2.5 mm wide; paleas concealed by lemmas.
    5. First glumes 1.3–3.5 mm long; palea apices with wide, V-shaped notches; stamens 3 (2).
        6. Lowest lemmas 2.0–2.5 mm long; anthers 0.5–0.8 mm long ...... *G. grandis*
        6. Lowest lemmas 2.5–3.5 mm long; anthers 1.0–2.0 mm long ...... *G. maxima*
    5. First glumes 0.6–1.5 mm long; palea apices with narrow slits; stamens 2.
        7. Ligules usually closed in front, those of the lower leaves mostly 1–3 mm long .... *G. striata*
        7. Ligules usually open in front, those of the lower leaves 3–6 mm long ...... *G. elata*

***Glyceria borealis* (Nash) Batch.**
Northern mannagrass
Habitat/Range: Swamps, lakeshores, fens and marshes (often in water up to 1 m deep) in the lowland and montane zones; infrequent in SW BC, frequent in SC BC; N to AK and YT, E to NF and S to ME, PA, IN, SD, NM, AZ and CA.

***Glyceria canadensis* (Michx.) Trin.**
Rattlesnake-grass
Habitat/Range: Wet ditches, marshes and shallow water in the lowland zone; infrequent in the lower Fraser valley, where probably introduced; E to NF and S to NC, TN, IN and IL.

***Glyceria elata* (Nash) M.E. Jones**
Tall mannagrass
Habitat/Range: Swamps, lakeshores, streamsides, bogs and moist to wet meadows and forests in the lowland and montane zones; common in SW BC, less common E of the Coast-Cascade Mountains N to 55°N; E to SW AB and S to MT, NM, AZ, NM and CA.

***Glyceria grandis* S. Wats. ex A. Gray**
Reed or American mannagrass
Habitat/Range: Swamps, lakeshores, streamsides and wet meadows in the lowland, steppe and montane zones; common in S BC, less frequent northward; N to AK, YT and NT, E to NF and S to TN, IL, NM, AZ and OR.

***Glyceria leptostachya* Buckl.**
Slender-spiked mannagrass
Habitat/Range: Salt marshes, swamps, lakeshores, streamsides and wet meadows in the lowland zone; rare in coastal BC; N to SE AK and S to CA.

***Glyceria maxima* (Hartm.) Holmb.**
Giant mannagrass
Habitat/Range: Lakeshores and wet meadows in the lowland zone; rare on Vancouver Island, known recently from only Somenos Lake; introduced from Eurasia.

***Glyceria occidentalis* (Piper) J.C. Nels.** (*G. septentrionalis* auct. non A.S. Hitchc.)
Western mannagrass
Habitat/Range: Marshes, lakeshores and wet meadows and forests in the lowland zone; rare in coastal BC; S to ID and N to CA.

***Glyceria striata* (Lam.) A.S. Hitchc.** (*G. striata* var. *stricta* [Scribn.] Fern.)
Fowl mannagrass
Habitat/Range: Bogs, lakeshores and moist to wet meadows in the lowland to subalpine zones; common in BC S of 55°N, infrequent northward, absent on the Queen Charlotte Islands; N to AK, YT and NT.

HELICTOTRICHON

***Helictotrichon hookeri* (Scribn.) Henrard** (*Avenochloa hookeri* [Scribn.] Holub)
Spike-oat
Habitat/Range: Dry sites in the montane and subalpine zones; rare in N BC; N to SW YT and SW NT, E to S MB, disjunct in SW PQ and S to MN, ND, MT and NM.

HIEROCHLOË

1. Lemmas of the 2 staminate flowers conspicuously awned from a deeply bifid apex; ligules consisting of half membrane and half fringed hairs .... *H. alpina*

1. Lemmas of the 2 staminate flowers awnless; ligules mostly membraneous, slightly erose-lacerate and ciliolate .... *H. odorata*

***Hierochloe alpina* (Sw.) Roem. & Schult.** (*Anthoxanthum monticolum* [Bigelow] Veldk. in Schouten & Veldk.)
Alpine sweetgrass
Habitat/Range: Mesic to dry meadows in the subalpine and alpine zones; common in N BC, infrequent S to 51 1/2°N in SW and SC BC; circumpolar, N to AK, YT and NT, E to SW AB, disjunct to PQ and NF and S to NY, NH, VT and NW MT; Greenland, Eurasia.

***Hierochloe odorata* (L.) Beauv.** (*Anthoxanthum nitens* [Weber] Schouten & Veldk., *H. arctica* Presl)
Common sweetgrass
Habitat/Range: Beaches, moist meadows, streambanks and forest openings from the lowland to subalpine zones; common in BC east of the Coast-Cascade Mountains, rare on Vancouver Island; circumpolar, N to AK, YT and NT, E to NF and S to NJ, PA, SD, NM, AZ, NV and OR; Greenland, Iceland, Eurasia.

HOLCUS

1. Plants rhizomatous; lemma awns curved or bent; culms glabrous except at the bearded nodes .... *H. mollis*

1. Plants caespitose; lemma awns hooked; culms usually pubescent above and below the nodes .... *H. lanatus*

***Holcus lanatus* L.**
Common velvet-grass, or Yorkshire fog
Habitat/Range: Lawns, fields, roadsides, railways and waste areas; common in coastal BC, rare elsewhere in BC; introduced from Eurasia.

***Holcus mollis* L.**
Creeping softgrass, or creeping velvet-grass
Habitat/Range: Lawns, roadsides and moist waste areas; rare in SW BC; introduced from Europe.

HORDEUM[29]

1. Glumes, at least most of them, flattened and ciliate above the base; plants annuals .... *H. murinum*
1. Glumes not flattened, glabrous or scabrous; plants annuals or perennials.
    2. Lateral spikelets sessile, their florets fertile .... *H. vulgare*
    2. Lateral spikelets pedicellate, their florets sterile.
        3. Pedicels of lateral spikelets scabrous but not hirsute; plants introduced annuals .... *H. marinum*
        3. Pedicels of lateral spikelets hirsute; plants native perennials.
            4. Glumes (11) 20–90 mm long; inflorescence (with awns) nearly or quite as broad as long .... *H. jubatum*
            4. Glumes 7–12 mm long; inflorescence much longer than broad .... *H. brachyantherum*

***Hordeum brachyantherum* Nevski** (*H. jubatum* var. *breviaristatum* [Nevski] Bowd., *H. nodosum* auct. non L.)
Meadow barley
Habitat/Range: Moist to mesic saline meadows, sandy beaches and rocky bluffs in the lowland zone of coastal BC and mesic meadows and forest openings in the montane zone of interior BC; common in coastal BC, infrequent elsewhere in BC north to 55°N; north to AK, YT and NT, disjunctly E to NF and S to CO, NM and CA.

***Hordeum jubatum* L.** (*H. x caespitosum* [Scribn.] A.S. Hitchc. = ssp. *x intermedium*)
Foxtail barley
Habitat/Range: Moist to dry tidal shorelines, lakeshores, meadows, forests, roadsides and waste places in the lowland, steppe and montane zones; frequent in S BC, infrequent northward; circumpolar, N to AK, YT, and NT, E to NF and S to TN, TX, NM, AZ, MX and CA; Eurasia, S America.
Notes: *Hordeum jubatum* hybridizes with several species in other genera of the Triticeae as well as with *H. brachyantherum*. These latter hybrids, which tend to resemble *H. jubatum* more than *H. brachyantherum* are called *H. jubatum* ssp. *x intermedium* Bowden. The two subspecies occurring in BC may be recognized as follows:

1. Glumes 33–90 mm long; lemma awns of the central florets 33–90 mm long ..... ssp. *jubatum*
1. Glumes 11–36 mm long; lemma awns of the central florets 7–36 mm long .... ssp. *x intermedium* Bowd.

***Hordeum marinum* Huds. ssp. *gussonianum* Thell.** (*H. geniculatum* All.)
Mediterranean barley
Habitat/Range: Disturbed sites and waste places; rare on S Vancouver Island; introduced from Europe.

***Hordeum murinum* L.** (*H. leporinum* Link = ssp. *leporinum*)
Wall barley
Habitat/Range: Roadsides and waste places; infrequent in SW BC (both ssp.) and SC BC (ssp. *murinum*); introduced from Europe.
Notes: Two subspecies occur in BC.

---

29 Contributed by M.E. Barkworth.

1. Central spikelet florets sessile or their rachilla less than half the length of the pedicels of the lateral spikelets .... ssp. *murinum*
1. Central spikelets florets on rachilla equal or subequal to the length of the pedicels of the lateral spikelets .... ssp. *leporinum* (Link) Asch. & Gräbn.

***Hordeum vulgare* L.**
Common barley
Habitat/Range: Dry roadsides in the montane zone; rare in SC BC; introduced from Europe.

KOELERIA

***Koeleria macrantha* [Ledeb.] J.A. Schultes f.,** (*K. gracilis* Pers., *K. cristata* Pers., *K. nitida* Nutt., *K. pyramidata* [Lam.] Beauv.)
Junegrass
Habitat/Range: Dry sites in the steppe to subalpine zones; common in SC and SE BC, less frequent in SW and N BC; circumpolar, N to AK, YT and SW NT, E to Labr. and S to MO, OH, DE, TX, and MX; Eurasia.
Notes: A general concensus has yet to be reached regarding which name is best for this species. For a discussion of the problems involved see Hitchcock *et al.* (1969:609) and Holmgren and Holmgren (1977:254).

LEERSIA

***Leersia oryzoides* (L.) Sw.**
Rice cutgrass
Habitat/Range: Swamps, lakeshores and wet ditches in the lowland zone; rare in SW BC (lower Fraser Valley); circumpolar, E to NB and NS and S to most of the US, Eurasia, S America.

LEPTOCHLOA

***Leptochloa fascicularis* (Lam.) A. Gray**
Sprangletop
Habitat/Range: Waste places in the lowland zone; rare in the Vancouver area; introduced from the E United States.

LEYMUS[30]

1. Glumes lanceolate, conspicuously veined and flat, at least at the base; plants of coastal BC .... *L. mollis*
1. Glumes awnlike, not conspicuously veined; plants of interior BC.
   2. Lemmas conspicuously pubescent; plants rhizomatous, 1.8–11 dm tall .... *L. innovatus*
   2. Lemmas glabrous or inconspicuously pubescent; plants usually cespitose, 7–22 dm tall .... *L. cinereus*

---

30 Contributed by M.E. Barkworth.

Leymus cinereus **(Scribn. & Merr.) A. Löve** (*Elymus cinereus* Scribn. & Merr., *E. piperi* Bowd.)
Giant wildrye
Habitat/Range: Dry, usually gravelly or sandy sites and road sides in the steppe and montane zones; frequent in SC BC, infrequent in SE BC; E to SK and S to MN, AR, NM, AZ and CA.

Leymus innovatus **(Beal) Pilger** (*Elymus innovatus* Beal)
Fuzzy-spiked wildrye
Habitat/Range: Mesic to dry meadows and forests in the montane zone; common in BC east of the Coast-Cascade Mountains; N to AK, YT and NT, E to MB and S to SD.

Leymus mollis **(Trin. in Spreng.) Pilger** (*Elymus arenarius* auct. non L., *E. mollis* Trin. in Spreng.)
Dune wildrye or dunegrass
Habitat/Range: Moist to mesic sandy or gravelly beaches and edges of shoreline forests in the lowland zone; common in coastal BC; amphiberingian, N to AK, YT and NT, E to NF and S to NY and CA; Greenland, E. Asia.
Notes: A hybrid between *Leymus mollis* and *Elymus glaucus*, known only from Ucluelet and Gold River, has been named *Leymus x uclueletensis* (Bowd.) Baum (*Elymus x uclueletensis* Bowd.). A second hybrid which is infrequent in SW BC, *Leymus x vancouverensis* (Vasey) Pilger (*Elymus vancouverensis* Vasey), is a hybrid between *Leymus mollis* and *L. triticoides* (Buckl.) Pilger. The latter is not known from BC.

LOLIUM

1. Lemmas awnless .... *L. perenne*
1. Lemmas awned.
    2. Glumes shorter than the spikelets .... *L. multiflorum*
    2. Glumes usually exceeding the spikelets .... *L. temulentum*

Lolium multiflorum **Lam.**
Italian ryegrass
Habitat/Range: Fields, roadsides and waste areas; infrequent in SW BC, rare elsewhere in BC; introduced from Europe.

Lolium perenne **L.**
Perennial ryegrass
Habitat/Range: Fields, roadsides and waste areas; frequent in SW and SC BC, introduced elsewhere in grass-seeding mixtures; introduced from Eurasia.

Lolium temulentum **L.** (*L. temulentum* var. *leptochaeton* A. Br.)
Bearded ryegrass
Habitat/Range: Fields, roadsides and waste areas; infrequent in S BC; introduced from Eurasia.

MELICA[31]

1. Lemmas awned.
    2. Awns 0.5–3 mm long; panicle branches appressed .... *M. harfordii*
    2. Awns 3–10 mm long; panicle branches widespread, even reflexed .... *M. smithii*
1. Lemmas unawned or awns inconspicuous.

31 Contributed by M.E. Barkworth.

3. Lemmas strongly tapering and acuminate, usually pubescent over the veins .... *M. subulata*
3. Lemmas acute, obtuse or rounded.
    4. Lemmas with hairs on the marginal veins clearly longer than those elsewhere; awns present but inconspicuous .... *M. harfordii*
    4. Lemmas glabrous or scabrous.
        5. Glumes less than 1/2 the length of the spikelets; corms not clustered, remaining attached to a section of rhizome .... *M. spectabilis*
        5. Glumes 1/2 to 3/4 length of the spikelets; corms clustered, not remaining attached to the section of rhizome .... *M. bulbosa*

***Melica bulbosa* Geyer ex Porter & Coult.**
Oniongrass
Habitat/Range: Mesic to dry slopes in the steppe to subalpine zones; rare in SC BC, known only from Lillooet and Manning Park; S to CO and CA.

***Melica harfordii* Boland.**
Harford's melic
Habitat/Range: Dry slopes and open forests in the lowland zone; infrequent in SW BC; S to CA.

***Melica smithii* (Porter) Vasey**
Smith's melic
Habitat/Range: Moist forests in the lowland and montane zones; rare in S and EC BC; E to SW AB and disjunctly to E ON and PQ and S to WY and OR.

***Melica spectabilis* Scribn.**
Purple oniongrass
Habitat/Range: Wet to dry meadows and open forests in the montane and subalpine zones; rare in S BC east of the Coast-Cascade Mountains; E to SW AB and S to CO and N CA.

***Melica subulata* (Griseb. in Ledeb.) Scribn.**
Alaska oniongrass
Habitat/Range: Moist to dry meadows and forests in the lowland and montane zones; common on Vancouver Island; the Queen Charlotte Islands and SC and SE BC; N to SE AK, E to SW AB and S to WY and CA; Chile.

MUHLENBERGIA[32]

1. Panicle open with spreading branches at maturity, 2.5–15 cm wide, usually 1/3 to 2/3 the total height of the plant; spikelets on long, slender, capillary pedicels .... *M. asperifolia*
1. Panicle narrow and ascending, compact, less than 2.5 cm wide, usually less than 1/3 the total height of the plant; spikelets subsessile on short pedicels.
    2. Lemmas not pilose or bearded at base (appressed hairs very inconspicuous); leaf blades 0.6–1.6 mm wide.
        3. Plants perennials; rhizomes present, scaly and creeping; culms nodulose, greater than 1 mm in diameter; anthers generally 1.2–1.5 mm long .... *M. richardsonis*
        3. Plants annuals; rhizomes absent; culms not nodulose less than 1 mm in diameter; anthers generally 0.5–1.1 mm long .... *M. filiformis*
    2. Lemmas pilose or bearded at base; leaf blades 2–6 mm wide.

---

32 Contributed by P.M. Peterson.

4. Hairs at base of lemma numerous, 2.0–3.5 mm long; lemmas awned, 4–10 mm long .... *M. andina*
4. Hairs at base of lemma less numerous, less than 1.5 mm long; lemmas awn-tipped.
   5. Glumes 1.8–4.0 mm long, about equal to or less than the length of the lemma; panicles with slightly spreading branches; anthers 0.3–0.5 mm long .... *M. mexicana*
   5. Glumes 4–7 mm long, 1 1/2 to 2 times longer than the length of the lemma; panicles contracted with tightly appressed branches; anthers 0.5–1.5 mm long.
      6. Internodes dull, puberulent; lemmas pilose along margins and base; ligules less than 0.6 mm long; anthers 0.8–1.5 mm long; moist bogs .... *M. glomerata*
      6. Internodes smooth and polished except near the apex; lemmas pilose at base only; ligules 0.6–1.5 mm long; anthers 0.4–0.7 mm long; dry, often rocky habitats .... *M. racemosa*

***Muhlenbergia andina* (Nutt.) A.S. Hitchc.**

Foxtail muhly

Habitat/Range: Wet to moist canyons, streambanks, meadows, and hot springs (in tufa) in the steppe and montane zones; rare in SE BC, known only from Fairmont Hot Springs; disjunctly E to MB and S to TX, NM and CA.

Notes: This species has been collected only once – in 1941.

***Muhlenbergia asperifolia* (Nees & Meyen) Parodi**

Alkali muhly, or scratchgrass

Habitat/Range: Wet to moist alkaline meadows, seeps, and hot springs in the steppe and montane zones; frequent in S BC east of the Coast-Cascade Mountains; E to ON and S to TX, MX and CA.

***Muhlenbergia filiformis* (Thurb.) Rydb.**

Slender muhly, or pull-up

Habitat/Range: Wet to moist meadows, seeps, and streambanks in montane and subalpine zones; infrequent in S BC, rare in NE BC; E to AB and S to SD, KS, NM, AZ, CA and NW MX.

***Muhlenbergia glomerata* (Willd.) Trin.**

Marsh muhly

Habitat/Range: Wet to moist meadows, streambanks, bogs, irrigation ditches, lake margins, and hot springs in the steppe and montane zones; infrequent in SC BC, rare in E BC; N to NT, E to NF and S to MD, VA, TX, AZ, NV, MX and NE OR.

Notes: This species has been studied cytologically and most populations appear to be diploid ($2n=20$) whereas two close relatives, *M. mexicana* and *M. racemosa*, are tetraploids ($2n=40$). Separation of these taxa is often problematic and major differences are emphasized in the key.

***Muhlenbergia mexicana* (L.) Trin.**

Wirestem muhly

Habitat/Range: Wet, moist or mesic slopes, road embankments, ditches, lake margins, hot springs, meadows, and waste areas near railroad tracks in the steppe and montane zones; infrequent in E BC; E to NS and NB and S to NC, OH, KS, NM and CA.

***Muhlenbergia racemosa* (Michx.) B.S.P.**

Satin grass

Habitat/Range: Moist to mesic slopes, open forests and rocky areas in the steppe and montane zones; rare in SC BC, known only from Pavilion Lake; E to ON and S to OK, TN, NM and N MX.

Muhlenbergia richardsonis **(Trin.) Rydb.**
Mat muhly
Habitat/Range: Moist to dry, often alkaline meadows, talus slopes, and gravel bars along rivers in the steppe and montane zones; common throughout BC, except absent on the Queen Charlotte Islands; N to S YT and S NT, E to PQ and NB and S to NE, MI, MN, AZ, CA and NW MX.

NASSELLA[33]

Nassella viridula **(Trin.) Barkw.** (*Stipa viridula* Trin.)
Green needlegrass
Habitat/Range: Mesic to dry slopes and open forests in the steppe and montane zones; infrequent in NE, SC and SE BC; N to SW NT, E to MB and S to KS, MN, NM and AZ.

ORYZOPSIS[33]

1. Blades of cauline leaves absent or less than 1 cm long; callus hairs soft and wavy .... *O. asperifolia*
1. Blades of cauline leaves well-developed, more than 1 cm long; callus hairs strigose or absent.
   2. Awns 1–2 mm long, readily deciduous, absent from most florets at maturity .... *O. pungens*
   2. Awns 3–15 mm long, persistent.
      3. Awns 8–15 mm long, twice geniculate .... *O. canadensis*
      3. Awns 3–8 mm long, straight or once geniculate.
         4. Panicle branches widespread; awns straight but twisted .... *O. micrantha*
         4. Panicle branches appressed to the rachis; awns once-geniculate .... *O. exigua*

Oryzopsis asperifolia **Michx.**
Rough-leaved ricegrass
Habitat/Range: Mesic to dry, gravelly or rocky forest openings in the montane zone; frequent in BC east of the Coast-Cascade Mountains; N to NW NT, E to NF and S to IL, IN, WV, ND, UT, NM, ID and WA.

Oryzopsis canadensis **(Poir.) Torr.**
Canada ryegrass
Habitat/Range: Dry slopes, grasslands and open forests in the montane zone; rare in NE BC, known only from Pouce Coupe; E to NF and S to NH, MN and WI.

Oryzopsis exigua **Thurb. in Torr.**
Little ricegrass
Habitat/Range: Dry talus slopes rocky ridges and open forests in the steppe and montane zones; infrequent in S BC; E to AB and S to CO, UT, NV and OR.

Oryzopsis micrantha **(Trin. & Rupr.) Thurb.**
Small-flowered ricegrass
Habitat/Range: Mesic to dry, rocky slopes and open forests in the steppe and montane zones; infrequent in SC, rare in SE BC; E to SW MB and S to OK, ND, NM, AZ and NV.

---

33 Contributed by M.E. Barkworth.

Oryzopsis pungens **(Torr.) A.S. Hitchc.**
Short-awned ricegrass
Habitat/Range: Moist to dry, rocky or sandy sites in the montane zone; frequent in BC east of the Coast-Cascade Mountains; N to YT and NT, E to NB and NS and S to NJ, NY, IL, IN, SD and CO.

PANICUM[34]

1. Second glume and sterile lemma short pilose all over (sometimes glabrous in *P. oligosanthes* var. *scribnerianum*); basal leaves distinctly different from the culm leaves, forming a winter rosette; plants passing from one growth phase to another, the so-called vernal phase characterized by simple stems and terminal panicles becoming transformed to the autumnal phase with branched culms and axillary, often cleistogamous panicles.
   2. Plants velvety-villose, greyish green; ligules more than 2.5 mm long; spikelets 1.5–2 mm long .... *P. occidentale*
   2. Plants copiously papillose-hispid; ligules less than 1.5 mm long 1 mm long; spikelets 2.7–3 mm long .... *P. oligosanthes*
1. Second glume and sterile lemma glabrous, or sometimes scaberulous only on the nerves towards the apex; basal leaves similar to the culm leaves, not forming a winter rosette; spikelets all fertile.
   3. First glume 0.5–1 mm long, about 1/4 the length of the spikelet, apex obtuse .... *P. dichotomiflorum*
   3. First glume 1–3 mm long, at least 1/2 the length of the spikelet, apex acute or acuminate.
      4. Panicles drooping, spikelets 4.5–6 mm long .... *P. miliaceum*
      4. Panicles erect, spikelets 2.5–3.5 mm long .... *P. capillare*

Panicum capillare **L.** (*P. capillare* var. *occidentale* Rydb.)
Common witchgrass
Habitat/Range: Moist to dry roadsides, railway embankments, gravelly slopes and fields in the lowland, steppe and montane zones; frequent in S BC north to 51°N; E to NS and S to most of US.

Panicum dichotomiflorum **Michx.**
Smooth witchgrass
Habitat/Range: Moist railway embankments and waste areas; rare in SW BC (Cloverdale); introduced from E and central N America.

Panicum miliaceum **L.**
Broom-corn millet
Habitat/Range: Fields, roadsides, shorelines and waste areas; infrequent in S BC; introduced from Eurasia.
Notes: Occasionally cultivated in our region and continuously reintroduced from commercial birdseed.

Panicum occidentale **Scribn.** (*Dicanthelium acuminatum* [Sw.] Gould & Clark, *D. lanuginosum* [Elliot] Gould, *Panicum lanuginosum* Elliot, *P. pacificum* Hitchc. & Chase, *P. thermale* Boland.)
Western witchgrass
Habitat/Range: Moist to dry shores, beaches, open woods, meadows and bogs in the lowland and montane zones; frequent in S BC to 53°N and occasionally much farther north in association with hot springs; E to PQ and NS and S throughout the US.

---

34 Contributed by J.G. Packer.

***Panicum oligosanthes* Schult. var. *scribnerianum* (Nash) Fern.** (*Dicanthelium oligosanthes* [Schult.] Gould var. *scribnerianum* [Nash] Gould)

Few-flowered witchgrass

Habitat/Range: Dry open sandy flats, gravelly knolls and rocky bluffs in the lowland, steppe and montane zones; frequent in S BC to 51°N except for the extreme SE; E to ON and S throughout the US.

PHALARIS

1. Plants rhizomatous perennials, 5–15 (20) dm tall; panicles 7–15 (25) cm long .... *P. arundinacea*
1. Plants annuals, 3–6 (10) dm tall; panicles 1.5–3 (4) cm long .... *P. canariensis*

***Phalaris arundinacea* L.**

Reed canarygrass

Habitat/Range: Wet meadows, ditches and lakeshores in all but the alpine zone; common in S BC, rare northward; circumpolar, N to AK, YT and NT, E to NF and S to NC, TX, NM, AZ and CA; Eurasia.

Notes: The variegated ornamental form (f. *variegata* [Parnell] Druce [var. *picta* L.]) sometimes occurs as an escape.

***Phalaris canariensis* L.**

Canarygrass

Habitat/Range: Roads and waste areas; rare in SW, SC and NE BC; introduced from N Africa and the Canary Islands.

PHLEUM

1. Panicles subglobose to short-cylindric, 1–4.5 cm long; culms caespitose or non-bulbous at the base; plants native at higher elevations .... *P. alpinum*
1. Panicles slender-cylindric, 4–11 (16) cm long; culms bulbous at the base; plant introduced at lower elevations .... *P. pratense*

***Phleum alpinum* L.** (*P. commutatum* Gaudin)

Alpine timothy

Habitat/Range: Moist to mesic meadows and streambanks from the montane to alpine zones; common throughout BC except rare on the Queen Charlotte Islands and Vancouver Island; circumpolar, N to AK, YT and NT, E to NF and S to NH, ME, NE, NM, AZ and MX, Greenland, Iceland, Eurasia, S America.

***Phleum pratense* L.**

Common timothy

Habitat/Range: Fields, roadsides, waste areas and disturbed sites; common in S BC, less frequent northward; introduced from Eurasia.

PHRAGMITES

***Phragmites australis* (Cav.) Trin. ex Steud.** (*P. communis* [L.] Trin.)

Common reed

Habitat/Range: Marshes, ponds, lakeshores and ditches in the lowland, steppe and montane zones; infrequent in S and NE BC; circumpolar, N to SW NT, E to NF and NS and S to IN, LA, TX and MX; Eurasia, Trinidad, C America.

PLEUROPOGON[35]

Pleuropogon refractus **(A. Gray) Benth. in Vasey**

Nodding semaphoregrass

Habitat/Range: Swamps, streambanks, lakeshores, wet meadows and alluvial forests in the lowland zone; rare on S Vancouver Island and the adjacent lower mainland; S to CA.

POA[36]

1. Spikelets producing leafy bulblets in place of normal florets.
   2. Plants cespitose, with bulbous culm bases; leaf blades soft, soon withering, mostly less than 1 mm wide .... *P. bulbosa*
   2. Plants rhizomatous, lacking bulbous culm bases; leaf blades mostly broader.
      3. Culm leaf blades less than 5 cm long, flat or folded, strict; panicles erect; plants less than 25 cm tall .... *P. pratensis*
      3. Culm leaf blades more than 5 cm long, flat, lax; panicles lax; plants more than 25 cm tall .... *P. laxiflora*
1. Spikelets not producing bulblets.
   4. Anthers 0.2 to 1.1 mm long, not abortive.
      5. Panicles contracted; sheaths of upper culm leaves 1/5 or less their length; lemmas lacking a web; plants high alpine and mostly less than 10 cm tall.
         6. Lemmas glabrous or very sparsely pilose on the base of the keel .... *P. lettermanii*
         6. Lemmas pubescent on the keel and marginal nerves .... *P. abbreviata*
      5. Panicles generally open; sheaths of upper culm leaves closed more than 1/4 their length or sometimes less; lemmas with or without a web; plants of various habitats and heights.
         7. Plants annuals.
            8. Panicle branches smooth; lemmas lacking a web on the callus, villous on the nerves and glabrous between them, or glabrous throughout; paleas villous on the keels, not at all scabrous; .
               9. Anthers minute, less than 0.4 mm long, round prior to dehiscing; lemma nerves crisply villous; rachilla of the terminal florets 1/2–3/4 as long as the lemma .... *P. infirma*
               9. Anthers 0.5–1.0 mm long, linear prior to dehiscing; lemmas nerves softly villous or glabrous; rachilla of the terminal florets generally less than 1/2 as long as the lemma .... *P. annua*
            8. Panicle branches quite scabrous, lemmas with or without a sparse web on the callus, sparsely puberulous over much of the surface; paleas scabrous on the upper portion of the keels, often villous below .... *P. howellii*
         7. Plants perennials.
            10. Sheaths of upper culm leaves closed over 1/4 their length.
               11. Sheaths of the upper culm leaves closed to near the top; lemmas glabrous except for a web on the callus .... *P. marcida*

---

[35] Contributed by M.E. Barkworth.
[36] Contributed by R.J. Soreng.

11. Sheaths of the upper culm leaves closed 1/4–2/3 their length; lemmas at least sparsely villous on the keel and marginal nerves and with a web (or without in *P. laxa*) on the callus.

12. Plants rhizomatous; first glumes mostly more than 1-nerved.

13. Panicles sparsely-flowered, with 1–2 branches at the lower nodes; uppermost culm leaf-blades mostly as long or longer than their sheaths .... *P. laxiflora*

13. Panicles closely-flowered, with mostly 3–5 branches at the lower nodes; uppermost culm leaf-blades much shorter than their sheaths .... *P. pratensis*

12. Plants cespitose; first glumes mostly 1-nerved.

14. Panicle branches up to 3 cm long, bearing 2–6 closely arranged spikelets; first glumes 3-nerved; lemmas without a web ..... *P. laxa*

14. Panicle branches usually longer than 3 cm long, bearing 1–several sparsely arranged spikelets; first glumes 1–3-nerved; lemmas with a distinct (although sometimes sparse) web.

15. Panicles mostly less than 7 cm long, the branches smooth or sparsely scabrous, terete; lemmas broadly lanceolate; first glumes 1–3-nerved, broadly lanceolate .... *P. paucispicula*

15. Panicles 4–15 cm long, the branches sparsely to densely scabrous on the angles; lemmas narrowly lanceolate; first glumes 1-nerved, subulate .... *P. leptocoma*

10. Sheaths of upper culm leaves closed less than 1/5–1/4 their length.

16. Plants more or less stoloniferous; panicles usually 10–30 cm long; the uppermost leaves attached in the middle to upper part of the culm; keels of lemmas tending to arch in the apical part, toward the rachillas; rachillas never pilose .... *P. palustris*

16. Plants uniformly cespitose; panicles usually 2–13 cm long; the uppermost leaves attached in the lower 1/3 of the culm; keels of lemmas continuing straight into the apex or little arched inward; rachillas sometimes pilose .... *P. glauca*

4. Anthers 1.2–4.5 mm long, or vestigial (minute white slips) or withered (apparently aborted before maturity).

17. First glumes narrowly 1-nerved, more or less sickle shaped; lemmas short villous on the keel, glabrous elsewhere; palea nerves smooth or minutely bumpy; short-lived weakly stoloniferous perennials with quite scabrous branches and lax leaves with long acute ligules .... *P. trivialis*

17. First glumes mostly 3-nerved and rarely sickle shaped (except in *P. pratensis*); lemmas glabrous or pubescent but not as above; palea nerves scabrous to villous, rarely almost smooth; habits various.

18. Plants rhizomatous or stoloniferous.

19. Culms nodes and internodes decidedly compressed, keeled, several nodes exposed; panicle branches finely and evenly scabrous on the angles .... *P. compressa*

19. Culms nodes and internodes round (or slightly compressed) in cross-section, nodes exposed or not; panicle branches smooth or scabrous.

20. Dioecious or subdioecious species of sandy coastal habitats; leaf blades involute, all alike in form, those of the sterile shoots minutely scabrous to hispidulous on and between the veins of the upper surface.

21. Lemmas 5–8 mm long; robust plants of shifting sands ... *P. macrantha*

21. Lemmas 2.5–3 mm long; slender plants of more stable sands .... *P. confinis*

20. Perfect flowered species or plants of inland habitats; the culm leaf blades flat or folded, upper surface smooth, or minutely scabrous only on the nerves, or scabrous to hispidulous on and between the veins.

22. Sheaths of upper culm leaves closed less than 1/4 their length.

23. Lemmas thin, membranous; callus of the lemmas with a crown of short straight hairs surrounding the base; rhizomatous plants of coastal salt marshes .... *P. eminens*

23. Lemmas firm; callus of the lemmas glabrous or with a single tuft of hairs arising below the keel; mostly stoloniferous inland plants .... *P. palustris*

22. Sheaths of upper culm leaves closed from over (1/4) 1/3 their length to near the top.

24. Callus glabrous on all lemmas (rarely with a very short dorsal tufts of fine hairs).

25. Sheaths of at least some basal leaves retrorsly pubescent to scabrous the collar glabrous, scabrous, or with hairs no longer than those of sheath; upper culm sheaths closed 1/3–2/3 (4/5) their length; blades of the sterile shoots often involute, usually densely and minutely hirsute above, on and between the veins; lemmas 3–6.5 mm long; flowers pistillate; plants from east of the Coast-Cascade Mountains .... *P. wheeleri*

25. Sheaths of the lower culm and lateral shoots glabrous or rarely pubescent, the collar and throat villous, the hairs (if any) longer than those of the sheath; upper culm sheaths closed 2/3 or more their length; blades of the non-flowering shoots elongate and flat, smooth or sparsely scabrous on but not between the veins; lemmas 2.5–3.5 mm long; flowers pistillate, all perfect, or mixed; plants from west of the Coast-Cascade Mountains .... *P. nervosa*

24. Callus with distinct single dorsal tufts of long villous hairs, or a crown of hairs around the base of the lemma, on at least the lower lemmas within spikelets.

26. Lemmas firmer, not membranous throughout; callus with a single tuft of hairs from the dorsal surface; plants of various sizes but not of coastal salt marshes.

27. Paleas villous at least on the lower portion of the keels, at least sparsely puberulent or spinulose between them; panicle branches mostly singular or in pairs; lemma villous on 5 nerves and more or less puberulous to villous between them, the exposed surface smooth .... *P. arctica*

27. Paleas scabrous, glabrous, or rarely short villous on the lower portion of the keels, usually glabrous between them; lower panicle branches in fascicles of (2) 3 to 5; lemma villous on the keel and marginal nerves, and infrequently on the intermediate nerves, minutely bumpy on the exposed surface .... *P. pratensis*

26. Lemmas thin, membranous; callus with a crown of short straight hairs surrounding the based of the lemma; robust plants of coastal salt marshes .... *P. eminens*

18. Plants cespitose, more or less densely tufted.

28. Flowers all pistillate (in Canadian material), the anthers vestigal (minute white slips 0.1–1.0 mm long, or sometimes longer in *P. epilis,* but clearly non-functional); sheaths closed over (1/4) 1/3 the length or more; leaf blades of lateral shoots at least, folded to involute and densely scabrous to hispidulous on and between the veins of the upper surface.

29. Lemmas prominently villous on at least the keel and marginal nerves; culm blades strongly reduced upwards, the uppermost one often vestigial, rigid ... . *P. fendleriana*

29. Lemmas glabrous or sparsely pilose on the keel in the lower 1/3 and the base of the marginal nerves; culm blades not regularly or sharply reduced upwards, the uppermost never vestigal, not rigid .... *P. cusickii*

28. Flowers normally perfect, the anthers mostly 1.2 mm long or longer (sometimes aborted late in the maturation process); upper culm sheaths closed less than 1/4 their length (except in *P. alpina*); leaf blades, if folded or involute, at most scabrous on the adaxial nerves, glabrous between them.

30. Spikelets broadly rounded at the base in side view; culm sheaths closed about 1/4 their length, basal sheaths strongly overlapping and persistent; panicles 2–5 cm long, the branches smooth; lemmas distinctly villous; callus glabrous .... *P. alpina*

30. Spikelets narrower, not broadly rounded at the base in side view; upper culm sheaths closed less than 1/4 their length, basal sheaths various; panicles 2–30 cm long, the branches usually scabrous; lemmas villous or not; callus glabrous or pubescent.

31. Callus of the lemmas with single dorsal tufts of hairs, or glabrous, appearing laterally pinched in on back; palea nerves scabrous; rachilla joints mostly less than 1 mm long; all initial shoots of the season flowering; lemmas with distinct golden to bronze-tinged band near the apex at maturity.

32. Ligules of all leaves less than 0.5 mm long; uppermost leaves attached above the middle of the culms; lemmas sharply acute, the keel continuing straight into the apex; rachillas often pilose .... *P. nemoralis*

32. Ligules of at least some leaves 0.5–4.0 mm long, or the uppermost culm leaves attached below the middle of the culm, or the lemmas broadly acute to obtuse, the keel continuing straight into the apex or arching inward toward the rachilla; rachillas pilose, hirtsutulous, finely tuberculate or smooth.

33. Plants tending to be stoloniferous; panicles mostly 13–30 cm long; the uppermost leaves attached in the middle to upper part of the culm; keels of lemmas tending to arch in the apical part toward the rachilla; rachillas never pilose .... *P. palustris*

33. Plants uniformly cespitose; panicles mostly 2–13 cm long; the uppermost leaves attached in the lower 1/3 of the culm; keels of the lemmas continuing straight into the apex or little arched inward; rachillas sometimes pilose .... *P. glauca*

31. Callus of the lemmas with a crown of straight hairs around the base of the lemmas, or glabrous, smoothly rounded across the back; palea nerves somewhat villous in the lower portion, and often between them; rachillas joints elongate mostly more than 1 mm long; sterile shoots frequent, developing concurrently with the flowering shoots (except in some *P. stenantha*); lemmas white-hyaline or slightly bronze-colored at the apex.

34. Lemmas glabrous, or at most sparsely crisply puberulous on the lower nerves, smooth or scaberulous over the body; ligules truncate to obtuse, mostly less than 2 mm long, firm, scabrous on the back .... *P. secunda*

34. Lemmas villous on the keel and marginal nerves or at least finely puberulent over the lower 1/2; ligules of at least the upper culm leaves acute to acuminate, more than 2 mm long (sometimes shorter in dwarf plants), smooth or sparsely scabrous on the back.

35. Panicles open at maturity, 7–17 cm long; lemmas distinctly keeled, broadly lanceolate in side view, prominently villous on the keel and marginal nerves, and sometimes the midnerves, the internerves glabrous or puberulous; callus often hairy; leaf blades (1.5) 2 to 3 mm broad; plants loosely tufted .... *P. stenantha*

35. Panicles open or contracted at maturity, 5–20 cm long; lemmas indistinctly keeled to rounded across the lower back, narrowly lanceolate in side view, villous only on the keel and marginal nerves to more or less evenly puberulent over the lower 1/2; callus hairs minute and sparse, or lacking; if the panicle open, then the basal blades mostly less than 1.5 mm broad and soon withering; plants tightly tufted .... *P. secunda*

Poa abbreviata **R. Br. ssp.** ***pattersonii*** **(Vasey) A. Löve, D. Löve and Kapoor** (*P. abbreviata* ssp. *jordalii* [Pors.] Hult., *P. pattersonii* Vasey)

Abbreviated bluegrass

Habitat Range: Fell fields and rocky slopes (often on limestone) in the alpine zone; rare in BC in and E of the Coast-Cascade Mountains; amphiberingian, N to AK, YT and NT, E to AB and S to CO and UT; E Asia.

Notes: This species was last collected in 1943 in the Rocky Mountains of SE BC (Soreng 1991).

Poa alpina **L.**

Alpine bluegrass

Habitat/Range: Moist to wet meadows and talus slopes in the subalpine and alpine zones; common throughout BC; circumboreal, N to AK, YT and NT, E to NF and NS, and S to OR, CO, UT and NM; Greenland, Eurasia.

Poa annua **L.**

Annual bluegrass

Habitat/Range: Waste places, lawns, gardens in the lowland and montane zones; common throughout BC, especially S and W BC; introduced from Eurasia.

Poa arctica **R. Br.** (*P. arctica* ssp. *longiculmis* Hult., *P. longipila*, and *P. williamsii* Hult. = ssp. *arctica*, *P. lanata* Scribn. & Merr. and *P. malacantha* Kom. = ssp. *lanata*)

Arctic bluegrass

Habitat/Range: Moist to wet meadows and stream sides in the subalpine and alpine zones; ssp. *arctica* - frequent throughout BC, ssp. *lanata* - locally frequent in NW BC (poorly defined elsewhere); ssp. *arctica* - circumboreal, N to AK, YT and NT, E to NS, S to WA and NM; Iceland, Eurasia; ssp. *lanata* - amphiberingian, N to AK, YT and NT; E Asia.

Notes: Two subspecies occur in B.C.

1. Spikelets mostly greater than 6 mm long; lower lemmas 4.5 to 6 mm long; culm leaf blades often flat and more than 2 mm wide .... ssp. *lanata* (Scribner & Merr.) Soreng

1. Spikelets less than 6 mm long; lower lemmas 3 to 4.5 mm long; culm leaf blades mostly folded and less than 2 mm wide .... ssp. *arctica*

Poa bulbosa **L.**

Bulbous bluegrass

Habitat/Range: Dry to mesic waste places in the lowland, steppe and montane zones; frequent in SW and SC BC; introduced from Eurasia.

Poa compressa **L.**

Canada bluegrass

Habitat/Range: Moist to wet waste places, roadsides and meadows in the lowland and montane zones; infrequent in S BC; introduced from Eurasia or possibly native, occurring throughout US and S Canada, N to AK.

Poa confinis **Vasey**

Beach bluegrass

Habitat/Range: Semi-stable sandy sites and beaches in the lowland zone; infrequent in coastal BC; S to N CA.

Poa cusickii **Vasey** (*P. epilis* Scribn., *P. purpurascens* Vasey var. *epilis* [Scribn.] M.E. Jones = ssp. *epilis*, *P. subaristata* Scribn. in Beal non Phil. = ssp. *pallida*, *P. purpurescens* Vasey non Spreng., *P. alpina* var. *purpurascens* Vasey = ssp. *purpurascens*)

Cusick's or skyline bluegrass

Habitat/Range: Dry to mesic meadows, talus slopes and ridges in the steppe and montane to lower alpine zones; ssp *epilis* - frequent throughout BC except absent in NW BC, ssp. *pallida* - infrequent in SW BC, ssp. *purpurascens* - frequent in SC and SW BC; ssp. *epilis* - E to AB and S to CO, UT and CA, ssp. *pallida* - N to S YT, E to MB and S to W ND, CO, UT and CA, ssp. *purpurascens* - S to CA.

Notes: Three subspecies occur in BC. These include two pistillate, apomictic subspecies (ssp. *epilis* and *purpurascens*) and a sexual, although often apomictic subspecies (ssp. *pallida*). The ssp. *cusickii* has yet to be collected in BC, but may be found in the future. It is distinguished from ssp. *pallida* by its longest lower panicle branches exceeding 1.7 mm in length.

1. Plants forming dense tufts, the tillers all intravaginal, without bladeless leaves below, the prophylls at the base of the new shoots 1–4 cm long; 0–1 culm nodes slightly exerted from their subtending sheaths at maturity; panicle branches scabrous .... ssp. *pallida* Soreng

1. Plants forming sparse tufts, the tillers both intra- and extravaginal, the latter with 1–several bladeless leaves below and the prophylls less than 0.5 mm long; 1–2 culm nodes well exerted from their subtending sheaths at maturity; panicle branches smooth or sparsely scabrous.

2. Lemmas of at least some lower florets of some spikelets very sparsely to moderately pubescent on the base of the keel and frequently with a few hairs on the callus, panicles loosely contracted and with less than 20 spikelets ... ssp. *purpurascens* (Vasey) Soreng
2. Lemmas of all florets smooth to scabrous; panicles tending to be compact, strict, and with more than 20 spikelets .... ssp. *epilis* (Scribn.) Weber

***Poa eminens* J. S. Presl in C.B. Presl** (*Arctopoa eminens* [J.S. Presl in C.B. Presl] Probat.)
Eminent bluegrass
Habitat/Range: Coastal salt marshes and gravelly beaches in the lowland zone; rare in coastal BC, known only from Kitimat and Douglas Channel; amphiberingian, N to AK, disjunct in PQ and NF; E Asia.
Notes: The type locality (Nootka Sound) is apparently incorrect due to a mistake in labelling (S.J. Darbyshire, pers. comm.).

***Poa fendleriana* (Steud.) Vasey** (*P. longiligula* Scribn. & T. Williams)
Fendler's bluegrass or Muttongrass
Habitat/Range: Dry slopes, talus, and ridges, in the steppe and lower montane zones; ssp. *fendleriana* - rare in SC BC, ssp. *longiligula* - infrequent in BC; E to MB and S to MX.
Notes: Two pistillate, apomictic subspecies occur in BC.

1. Ligules more than 1.5 mm long, decurrent on the margins of the sheaths .... ssp. *longiligula* (Scribn. and T. Williams) Soreng
1. Ligules less than 1.5 mm long, not decurrent .... ssp. *fendleriana*

***Poa glauca* Vahl** (*P. rupicola* Nash = ssp. *rupicola*, *P. interior* Rydb. pro parte)
Glaucous bluegrass
Habitat/Range: Dry ridges, slopes, stony meadows, and mossy ledges in the montane to alpine zones, common throughout BC; circumboreal, E to NF and S to NM and CA; Iceland, Eurasia.
Notes: Two subspecies occur in BC.

1. Lemmas puberulent between the keel and marginal nerves, callus of all florets entirely glabrous; dwarf alpine plants .... ssp. *rupicola* (Nash) W.A. Weber
1. Lemmas variously puberulent or pubescent, calluses webbed or not; if dwarf alpine plants, then callus of at least the basal florets within a spikelet with at least a vestige of a web or the lemmas glabrous between the keel and marginal nerves .... ssp. *glauca*

Plants that are green (rather than glaucous) with lemmas short villous on the keel and marginal nerves belong to the form, known as *P. interior* Rydb. This entire species complex might better be placed under *P. nemoralis.* In the alpine zone of NW BC *Poa pseudoabbreviata* Roshev. is likely to occur (Cody *et al.* 1990). It is distinguished from the above by its small tufts of fine leaves, culms less than 10 cm tall and sparsely flowered open panicles with scabrous branches.

***Poa howellii* Vasey & Scribn.**
Howell's bluegrass
Habitat/Range: Mesic meadows and dry oak forests in the lowland zone; infrequent in extreme SW BC; S to CA.

***Poa infirma* Kunth**
Diploid annual bluegrass
Habitat/Range: Mesic sites in the lowland zone; rare in SW BC, known only from Trial Island; introduced from S. Europe.

***Poa laxa* Haenke ssp. *banfianna* Soreng**
Banff bluegrass
Habitat/Range: Moist meadows and ridges in the alpine zone; rare in SE BC, known only from Yoho National Park.
Notes: This species has not been collected in BC since 1943 (Soreng 1991).

Poa laxiflora Buckl.
Lax-flowered bluegrass
Habitat/Range: Moist old-growth forests and rocky slopes in the lowland and montane zones, infrequent in and W of the Coast-Cascade Mountains; N to SE AK and S to NW OR.

Poa leptocoma Trin.
Bog bluegrass
Habitat/Range: Wet to moist meadows, streams and bogs in the subalpine and alpine zones; frequent in BC except in NE BC; amphiberingian, N to extreme SW YT and S AK and S to CO, UT, NM and CA; E Asia.

Poa lettermanii Vasey
Letterman's bluegrass
Habitat/Range: Mesic to dry rocky slopes and meadows in the subalpine and alpine zones; infrequent throughout BC east of the Coast Mountains; S to CO, UT and CA.
Notes: *Poa suksdorfii* (Beal) Vasey ex Piper, which occurs in N WA, may be expected in the Cascades of SW BC. It will key out to *P. lettermanii* but differs in its longer lemmas (4.1–5.8 mm) and anthers (0.8–1.7 mm).

Poa macrantha Vasey (*P. douglasii* Nees. ssp. *macrantha* [Vasey] Keck)
Seashore or sand-dune bluegrass
Habitat/Range: Shifting sands (especially dunes) in the lowland zone; infrequent in coastal BC; N SE AK and S to N CA.

Poa marcida A. Hitchc.
Weeping bluegrass
Habitat/Range: Wet to moist meadows, swamps and alluvial forests in the lowland zone; infrequent on Vancouver Island; rare on the lower mainland (Vancouver); S to NW OR.

Poa nemoralis L. (*P. interior* Rydb. pro parte)
Wood bluegrass
Habitat/Range: Moist forests and disturbed sites in the lowland zone; infrequent in SW BC; introduced from Eurasia.
Notes: This species perhaps should include *P. glauca* since nearly continuous variation occurs between them with *P. interior* forms being intermediate.

Poa nervosa (Hook.) Vasey
Coastal bluegrass
Habitat/Range: Moist forest slopes and mossy ledges along streams in the lowland and montane zones; rare in SW BC, known only from Nootka Sound; S to NW OR.
Notes: In BC this species is known only from the type collection at Nootka Sound.

Poa palustris L.
Fowl bluegrass
Habitat/Range: Wet to moist disturbed meadows and riparian sites in the lowland and montane zones; frequent in S and W BC; circumboreal, N to AK, YT and NT, E to NF and S to TN, NC, MO, UT, AZ and CA; Eurasia.

Poa paucispicula Scribner (*P. leptocoma* ssp. *paucispicula* [Scribner] Tzvelev, *P. merrilliana* Scribner & Merr. non Stapf., *P. glacialis* A. Hitchc.)
Western bluegrass
Habitat/Range: Mesic fellfields and meadows in the alpine zone; frequent throughout BC; amphiberingian, N to AK, YK and NT, E to AB and S to MT, NE and N WA; Asia.

***Poa pratensis* L.** (*P. agassizensis* Boivin & D. Löve, *P. alpigena* [Blytt] Lindm., *P. angustifolia* L., *P. irrigata* Lindm., *P. rigens* Hartm., *P. subcaerulea* Smith)

Kentucky bluegrass

Habitat/Range: Meadows and disturbed sites from the lowland to alpine zones; infrequent to common throughout BC; introduced from Eurasia as a lawn and pasture grass (ssp. *pratensis,* ssp. *irrigata* and ssp. *angustifolia*) or native (ssp. *agassizensis,* ssp. *alpigena,* and ssp. *colpodea*).

Note: Although the major forms in the *P. pratensis* complex are distinguished as species in Europe or subspecies in Eurasia, due to an abundance of information they are extremely difficult to separate in temperate N America. The key to extreme forms is presented here so that interested readers may compare concepts among varying authors and continue the study of the taxa. At least six subtly distinct facultatively apomictic subspecies occur in BC.

1. Flowers forming bulblets, panicles more or less linear, native in the subalpine and alpine zones of NW BC .... ssp. *colpodea* (Th. Fries) Tzvelev

1. Flowers usually not forming bulblets, panicles broader.

    2. Plants usually less than 25 cm tall, somewhat covered with a bluish coating; panicle branches arising from lower nodes in groups of 1–2 (3); introduced in coastal BC .... ssp. *irrigata* (Lindm.) Lindb. f.

    2. Plants 20–100 cm tall, larger plants without a bluish coating; panicle branches arising from lower nodes in groups of 3–5; ranges various.

        3. Blades of the vegetative shoots involute, less than 0.7 mm in diameter, elongated, 10–40 cm long, distinctly narrower then those of the flowering shoots; panicle branches smooth or sparsely scabrous; sporadically introduced in SW BC .... ssp. *angustifolia* (L.) Arcang.

        3. Blades of the sterile shoots flat or folded and involute on the margins, mostly more than 0.8 mm in diameter, not differentiated from those of the flowering shoots in form; panicle branches smooth or scabrous.

            4. Panicles 3–6 cm long, contracted at maturity, blades of sterile shoots 3–15 cm long, firm, folded and involute on the margins, often with sparse soft hairs above (also present in ssp. *angustifolia*); native of semiarid grasslands and forests in the steppe and montane SE BC .... ssp. *agassizensis* (Boivin & D. Löve) Taylor & MacBryde

            4. Panicles 4–15 cm long, open at maturity, blades flat or folded, mostly glabrous above.

                5. Spikelets lanceolate, strongly purple-colored; lemmas 2–3.5 mm long, villous on the intermediate nerves or glabrous; palea scabrous or villous on the lower portion of the keels; panicles narrowly pyramidal, branches smooth; leaf blades of the vegetative shoots folded; native in the subalpine and alpine zones of N BC .... ssp. *alpigena* (Blytt) Hitt.

                5. Spikelets broadly lanceolate to ovate, green or somewhat purple-colored; lemmas 2.5–3.5 mm long, glabrous between the keel and marginal nerves; palea scabrous on the keels; panicles pyramidal, the branches more or less scabrous; leaf blades of the vegetative shoots flat or folded; widely introduced in BC .... ssp. *pratensis*

***Poa secunda* J. S. Presl in C.B. Presl** (*P. ampla* Merr., *P. juncifolia* Scribn. and *P. nevadensis* Vasey = ssp. *juncifolia*; *P. canbyi* [Scribn.] Howell, *P. gracillema* Vasey, *P. incurva* Scribn., *P. sandbergii* Vasey and *P. scabrella* [Thurb.] Benth. = ssp. *secunda*)

Sandberg or Nevada bluegrass

Habitat/Range: Dry to moist ground from the steppe to lower alpine zones; ssp. *juncifolia* - common in saline to alkaline meadows and grasslands to rich meadows in the steppe and montane zones; ssp. *secunda* - common on well-drained meadows and grasslands in the steppe and montane zones; N to extreme S AK and YK, disjunct in E Canada, and S to MX.

Notes: Two variable, facultatively apomictic subspecies occur in BC.

1. Lemmas glabrous or scaberulous, at most crisp-puberulent on the nerves near the base; palea nerves scabrous; ligules all less than 2 mm long truncate to obtuse, firm, scabrous on back .... ssp. *juncifolia* (Scribn.) Soreng

1. Lemmas puberulous to villous on the nerves and between them, rarely glabrous between the nerves; palea mostly villous on the nerves below and between them; ligules of at least the upper culm leaves more than 2 mm long, acute to acuminate, smooth or scabrous .... ssp. *secunda*

A hybrid, *P. secunda* ssp. *juncifolia* x *P. pratensis* (*P. fibrata),* is rare in alkaline meadows of SC BC.

***Poa stenantha* Trin.**

Narrow-flowered bluegrass

Habitat/Range: Moist to mesic meadows and forest openings in the lowland to alpine zones; common in W and S BC; amphiberingian, N to AK and NT, E to AB and S to CO and OR; E Asia.

***Poa trivialis* L.**

Rough bluegrass

Habitat/Range: Moist disturbed sites in the lowland zone, infrequent in SW BC, rare elsewhere; introduced from Eurasia.

***Poa wheeleri* Vasey** (*P. nervosa* var. *wheeleri* [Vasey] C.L. Hitchc.)

Wheeler's bluegrass

Habitat/Range: Mesic meadows and open forests in the montane and subalpine zones; frequent E of the Coast-Cascade Mountains N to 57°N; E to MB and S to NM and CA.

POLYPOGON

***Polypogon monspeliensis* (L.) Desf.**

Rabbitfoot polypogon

Habitat/Range: Wet to dry waste areas and vernal pools; infrequent in S BC; introduced from Europe.

PUCCINELLIA[37]

1. Lemma margins entire throughout; pedicels glabrous, or nearly glabrous and bearing a few scattered scabrules; plants of coastal sites .... *P. pumila*

1. Lemma margins entire below, scabrous-serrate near the apices; pedicels usually uniformly scabrous, occasionally nearly glabrous and bearing a few scattered scabrules; plants of coastal and interior sites.

---

37 Contributed by J.I. Davis.

2. Lemma apices broadly obtuse to truncate, the first lemma less than 2.0 mm long; lower branches of fruiting inflorescences usually descending, occasionally spreading; plants mainly of interior sites .... *P. distans*

2. Lemma apices acute to broadly obtuse, the first lemma 2 mm long or longer; lower branches of fruiting inflorescences erect to spreading, occasionally descending; plants of coastal and interior sites.

   3. Lower branches of fruiting inflorescences usually erect to ascending, occasionally spreading; first lemma (2.2) 2.5–3.5 (4.1) mm long; plants of coastal sites .... *P. nutkaensis*

   3. Lower branches of fruiting inflorescences ascending to spreading, sometimes reflexed; first lemma 2.0–2.5 (3.0) mm long; plants predominantly of interior sites, occasionally of coastal sites .... *P. nuttalliana*

Puccinellia distans **(Jacq.) Parl.**

Weeping or European alkaligrass

Habitat/Range: Saline flats, lakeshores, coastal strands and disturbed sites in the lowland, montane and steppe zones; frequent throughout BC; introduced from Eurasia.

Notes: Generally recognized as introduced but some populations with branches of the fruiting inflorescences nondescending, including some in BC, are possibly intergradational with *P. nuttalliana*, and may also be native.

Puccinellia nutkaensis **(J.S. Presl in C.B. Presl) Fern. & Weath.** (*P. grandis* Swallen, *P. lucida* Fern. & Weath)

Pacific or Alaska alkaligrass

Habitat/Range: Beaches and marshes in the lowland zone; frequent in coastal BC; N to AK and S to CA, also along the Atlantic Coast of Canada and the NE US.

Notes: *Puccinellia nutkaensis* is barely distinct from *P. nuttalliana*, and like other named entities may represent an ecological race of the latter species. Larger plants in E and W North America have been recognized as *P. lucida* Fern. & Weath., but they are not consistently differentiable from *P. nutkaensis*.

Puccinellia nuttalliana **(Schult.) A.S. Hitchc.** (*P. airoides* [Nutt.] S. Wats. & J. Coult., *P. agrostidea* T. Sorensen, *P. borealis* Swallen, *P. deschampsiodes* T. Sorensen, *P. interior* T. Sorensen)

Nuttall's alkaligrass

Habitat/Range: Moist (sometimes ephemeral), usually saline sites in the lowland, montane and steppe zones; frequent in BC east of the Coast-Cascade Mountains, rare in coastal BC; N to AK, YT and NT, E to MB and S to WI and CA.

Notes: Many variants have been recognized in this complex on the basis of anther length, inflorescence branching pattern and other characters, but additional work will be required before recognition of additional species is justified. The variant forms appear to be merely weakly differentiated races connected to one another through intermediate forms.

Puccinellia pumila **(Vasey) A. S. Hitchc.**

Dwarf alkaligrass

Habitat/Range: Coastal beaches and marshes in the lowland zone; frequent in coastal BC; N to AK and S to CA, also along the Atlantic Coast of Canada and the northeastern US.

SCHIZACHNE

Schizachne purpurascens **(Torr.) Swallen**

False melic

Habitat/Range: Moist to dry open forests and rocky slopes in the montane zone; frequent in C and SC BC, rare in SE BC; circumpolar, N to AK, YT and NT, E to NF and S to WI, PA, SD, NM and MX; Eurasia.

SCHIZACHYRIUM

***Schizachyrium scoparium* (Michx.) Nash in Small** (*Andropogon scoparius* Michx.)
Little bluestem
Habitat/Range: Dry sites in the steppe and lower montane zones; rare in SE BC, known recently only from Bull River and Fairmont Hot Springs; E to NS and S to most of the U.S., MX.

SCOLOCHLOA

***Scolochloa festucacea* (Willd.) Link** (*Fluminea festucacea* [Willd.] A.S. Hitchc.)
Rivergrass
Habitat/Range: Ponds, marshes, lakeshores and streamsides in the steppe and montane zones; rare in EC, SC, and SE BC; N to NT, E to MB and S to IA, NB, MN, SD, MT and OR.

SECALE[38]

***Secale cereale* L.**
Rye
Habitat/Range: Disturbed sites and roadsides; rare and not usually persisting in S BC in and E of the Coast-Cascade Mountains; introduced from Eurasia.

SETARIA

1. Bristles downwardly barbed; panicles interrupted .... *S. verticillata*
1. Bristles upwardly barbed; panicles not interrupted.
    2. Spikelets subtended by (4) 5–6 (20) bristles; fertile lemmas distinctly cross-wrinkled ... *S. glauca*
    2. Spikelets subtended by (1) 2–3 (4) bristles; fertile lemmas not distinctly cross-wrinkled.
        3. Fertile lemmas and paleas closely surrounded by the sterile lemma and second glume, not shed separately from them .... *S. viridis*
        3. Fertile lemmas and paleas loosely surrounded by the sterile lemma and second glume, tending to be shed separately from them .... *S. italica*

***Setaria glauca* (L.) Beauv.** (*S. lutescens* [Weigel] Hubb.)
Yellow bristlegrass
Habitat/Range: Fields, roadsides and waste areas; infrequent in SW and SC BC; introduced from Eurasia.

***Setaria italica* (L.) Beauv.**
Foxtail millet
Habitat/Range: Fields, roadsides and waste areas; rare in SW BC; introduced from Eurasia.
Notes: Closely related to *S. viridis.*

***Setaria verticillata* (L.) Beauv.** (*S. verticillata* f. *ambigua* [Guss.] Boivin)
Bur bristlegrass
Habitat/Range: Fields, roadsides and waste areas; rare in S BC; introduced from Eurasia.

***Setaria viridis* (L.) Beauv.**
Green bristlegrass
Habitat/Range: Fields, roadsides and waste areas; frequent in S BC; introduced from Eurasia.
Notes: Closely related to *S. italica*.

---

38 Contributed by M.E. Barkworth.

SPARTINA

1. Spikes appressed; glumes and lemmas ciliate on the keels .... *S. gracilis*
1. Spikes ascending to spreading; glumes and lemmas scabrous on the keels .... *S. patens*

***Spartina gracilis* Trin.**
Alkali cordgrass
Habitat/Range: Marshes, wet ditches and meadows to dry (often saline) sites in the steppe and montane zones; frequent in SC and SE BC; E to SW ON and S to KS, NE, NM, AZ and CA.

***Spartina patens* (Ait.) Muhl.**
Salt meadowgrass
Habitat/Range: Saltmarshes in the lowland zone; rare in SW BC, known only from Comox and Vancouver; introduced from E North America.

SPHENOPHOLIS

***Sphenopholis obtusata* (Michx.) Scribn.** (*S. intermedia* [Rydb.] Rydb. = var. *major*)
Prairie wedgegrass
Habitat/Range: Moist meadows and streambanks, shallow ponds and hot springs in the steppe and montane zones; rare in NE (var. *major*) and SC and SE BC (var. *obtusata*); N to AK, E to NB and NS and S to most of the U.S.
Notes: This variable taxon is apparently represented by two varieties in BC.

1. Upper glumes rounded at the tips; inflorescences relatively dense .... var. *obtusata*
1. Upper glumes obtuse to acute at the tips; inflorescences paniculate .... var. *major* (Torr.) Erdman

SPOROBOLUS

1. Plants annuals; panicles usually concealed by the swollen sheaths .... *S. vaginiflorus*
1. Plants perennials; panicles projecting beyond sheaths.
   2. Panicles open, branches spreading; stems terete, pithy, sometimes hollow .... *S. airoides*
   2. Panicles spikelike, branches erect, if panicles slightly open then branches ascending or spreading; stems solid, sulcate.
      3. Spikelets 3.5 mm long, or more; branches erect .... *S. asper*
      3. Spikelets 3.0 mm long, or less; branches ascending .... *S. cryptandrus*

***Sporobolus airoides* (Torr.) Torr.**
Hairgrass dropseed
Habitat/Range: Moist to mesic sites in the steppe zone; rare in SC BC, known only from Osoyoos and Keremeos; S to SD, TX, MX and CA.

***Sporobolus asper* (Michx.) Kunth**
Rough dropseed
Habitat/Range: Dry sites in the steppe zone; rare in SC BC; E to NB and S to LA, TN, TX, NM and AZ.

***Sporobolus cryptandrus* (Torr.) A. Gray** (*S. cryptandrus* ssp. *typicus* A. Gray var. *occidentalis* E.K. Jones & Fassett, *S. cryptandrus* ssp. *fuscicolus* [Hook.] E.K. Jones & Fassett)
Sand dropseed
Habitat/Range: Mesic to dry sites in the steppe and montane zones; common in SC BC, infrequent elsewhere in S BC; E to SW PQ and S to IA, MO, MN, TX, NM and CA.

Sporobolus vaginiflorus (Torr.) Wood
Poverty grass
Habitat/Range: Waste places in the lowland zone; rare in the Vancouver area; introduced from the E United States.

STIPA[39]

1. Awns 5–20 cm long; callus 2.5–5 mm long, sharply pointed.
    2. Lemmas evenly pubescent at maturity; terminal segment of awns usually flexuous; ligules of basal leaves thin .... *S. comata*
    2. Lemmas with lines of pubescence or wholly glabrous at maturity; terminal segments of awns straight; ligules of basal leaves thick.
        3. Lemmas 12–16 mm long; awns 10–17 mm long; lower nodes with lines of pubescence .... *S. spartea*
        3. Lemmas 7.5–10 mm long; awns 6–9 mm long; lower nodes glabrous .... *S. curtiseta*
1. Awns 0.3–6 cm long; callus 0.3–3 mm long, often blunt.
    4. Awns 0.3–0.6 cm long, deciduous; lemmas long pubescent, the hairs to 6 mm long .... *S. hymenoides*
    4. Awns 1.6–6 cm long, persistent; lemmas pubescent with hairs about 1 mm long.
        5. First two segments of the awns pubescent with hairs about 1 mm long .... *S. occidentalis*
        5. First two segments of the awns scabrous or with hairs less than 0.5 mm long.
            6. Panicle branches flexuous and spreading .... *S. richardsonii*
            6. Panicle branches straight, appressed or ascending.
                7. Florets terete; paleas about half as long as the lemmas; lemma apices with minute membranous lobes .... *S. nelsonii*
                7. Florets somewhat laterally compressed; paleas as long as the lemmas; lemma apices with thick lobes .... *S. lemmonii*

Stipa comata Trin. & Rupr. (*S. comata* var. *intermedia* Scribn. & Tweedy, *S. tweedyi* Scribn.)
Needle-and-thread grass
Habitat/Range: Dry, open sites in the steppe and montane zones; common in SC and SE BC; N to YT, E to ON and S to MI, IA, IN, NE, TX, NM, AZ and CA.

Stipa curtiseta (A.S. Hitchc.) Barkw. (*S. spartea* Trin. var. *curtiseta* A.S. Hitchc.)
Short-awned porcupinegrass
Habitat/Range: Dry slopes in the steppe zone; frequent in SC and SE BC; N to NT, E to MB and S to SD, MT and WY.

Stipa hymenoides Roem. & Schult. (*Oryzopsis hymenoides* [Roem. & Schult.] Ricker in Piper)
Indian ricegrass
Habitat/Range: Dry slopes in the steppe zone; frequent in SC and SE BC; E to SW MB and S to TX, NM, AZ, CA and MX.

Stipa lemmonii (Vasey) Scribn.
Lemmon's needlegrass
Habitat/Range: Dry, open slopes and rocky bluffs in the lowland and montane zones; rare in SW BC; S to UT, NV and CA.

---

39 Contributed by M.E. Barkworth.

Stipa nelsonii **Scribn. var.** ***dorei*** **Barkw. & Maze** (*S. columbiana* auct. non Macoun, *S. occidentalis* var. *minor* [Vasey] C.L. Hitchc.)
Columbian needlegrass
Habitat/Range: Mesic to dry slopes and open forests from the lowland to subalpine zones; frequent in NE, SC and SE BC; N to YT, E to W MB and S to UT, AZ and CA.

Stipa occidentalis **Thurb. ex. S. Wats. var.** ***pubescens*** **Maze, Taylor & MacBryde** (*S. elmeri* Piper & Brodie ex Scribn.)
Stiff needlegrass
Habitat/Range: Dry slopes and open forests in the steppe and montane zones; frequent in NE, SC and SE BC; S to WY, UT, and CA.

Stipa richardsonii **Link**
Spreading needlegrass
Habitat/Range: Moist to dry slopes and open forests in the steppe and montane zones; frequent in SC and SE BC; N to AK and YT, E to MB and S to IA, CO, ID and WA.

Stipa spartea **Trin.**
Porcupinegrass
Habitat/Range: Dry to mesic slopes and open forests in the steppe and montane zones; rare in SC BC; E to ON and S to PA, IA, IN, OK and NM.

TORREYOCHLOA[40]

1. Lemmas with 7 distinct nerves, 2.5–3.5 mm long; spikelets 6–7 mm long .... *T. pallida*
1. Lemmas with 5 distinct nerves (sometimes with 2 indistinct marginal ones), 2.0–2.5 mm long; spikelets 4–5 mm long .... *T. pauciflora*

Torreyochloa pallida **(Torr.) Church** (*Glyceria fernaldii* [A.S. Hitchc.] St. John, *G. pallida* Torr.)
Fernald's false-manna
Habitat/Range: Shallow water, marshes, bogs and wet meadows in the montane zone; rare in EC and NE BC; amphiberingian, E to NF and S to MO, TN and VA; E. Asia.

Torreyochloa pauciflora **(J.S. Presl in C.B. Presl) Church** (*Glyceria pauciflora* J.S. Presl in C.B. Presl; *Puccinellia pauciflora* [J.S. Presl in C.B. Presl] Munz)
Weak false-manna or weak alkaligrass
Habitat/Range: Shallow water, marshes, bogs and wet meadows in the lowland to subalpine zones; frequent throughout BC; N to AK, E to AB and S to SD, CO, NM and CA.

TRISETUM

1. Lemmas awnless or with awns rarely exceeding the lemmas .... *T. wolfii*
1. Lemmas awned, the awns far exceeding the lemmas.
    2. Panicles spikelike, dense; basal sheaths puberulent; upper glumes not much longer than lower ones .... *T. spicatum*
    2. Panicles open to narrow (but not spikelike), loose; basal sheaths glabrous; upper glumes much longer than lower ones .... *T. cernuum*

---

[40] Contributed by J.I. Davis.

***Trisetum cernuum* Trin.** (*T. canescens* Buckl. = var. *canescens*)
Tall trisetum (var. *canescens*), nodding trisetum (var. *cernuum*)
Habitat/Range: Moist to dry sites in the lowland, steppe and montane zones; var. *canescens* - frequent in SW BC, infrequent in SC BC, var. *cernuum* - common in coastal BC, less frequent eastward; var. *canescens* - S to ID, MT and CA; var. *cernuum* - N to S AK, E to SW AB and S to ID, MT and N CA.
Notes: Two varieties occur in BC.

1. Panicles open, lax or drooping; leaves 6–11 mm wide .... var. *cernuum*
1. Panciles narrow, ascending; leaves 4–7 mm wide .... var. *canescens* (Buckl.) Beal

***Trisetum spicatum* (L.) Richt.** (*T. spicatum* ssp. *alaskanum* [Nash] Hult., ssp. *majus* [Vasey] Hult. and ssp. *molle* [Michx.] Hult.)
Spike trisetum
Habitat/Range: Moist to dry, often rocky sites in all vegetation zones; common throughout BC although less frequent along the coast; circumpolar, N to AK, YT and NT, E to NF and S to NH, NC, MN, PA, NM, AZ, CA and MX; Eurasia, S. America.
Notes: This is a highly variable species which cannot realistically be separated into infraspecific taxa. Even Hulten (1959), who recognized 14 subspecies in the *T. spicatum* complex, stated that ... "the differentiation between the different taxa becomes somewhat arbitrary."... for our western cordilleran material.

***Trisetum wolfii* Vasey**
Wolf's trisetum
Habitat/Range: Wet meadows and streamsides in the subalpine zones; rare in SE BC (known only from Mt. Assiniboine Provincial Park); E to SK and S to MT, NM and CA.

TRITICUM [41]

***Triticum aestivum* L.**
Wheat
Habitat/Range: Waste places and roadsides; rare escape in S BC; introduced from Eurasia.

VAHLODEA

***Vahlodea atropurpurea* (Wahlenb.) Fries in Hartm.** (*Deschampsia atropurpurea* [Wahlenb.] Scheele, *D. atropurpurea* var. *latifolia* [Hook.] Scribn. ex Macoun, *Vahlodea atropurpurea* ssp. *paramushierensis* [Kudo] Hult.)
Mountain hairgrass
Habitat/Range: Moist to mesic meadows, streambanks and lake margins in the montane to alpine zones; common in S BC, less frequent northward; circumpolar, N to AK, S YT and SW NT, E to Labr. and W NF and S to NY, NH, CO and CA; Greenland, N Eurasia.

VENTENATA [42]

***Ventenata dubia* (Leers) Coss. & Dur.**
Ventenata
Habitat/Range: Dry roadsides in the coastal lowlands; rare in SW BC, known only from Coquitlam; introduced from S Eurasia.

---

41 Contributed by M.E. Barkworth.
42 Contributed by W.J. Crins.

VULPIA[43]

1. First glumes rudimentary, less than 1/2 the length of the second .... *V. myuros*
1. First glumes not rudimentary, 1/2 or more the length of the second.
    2. Spikelets with 4–11 (–17) florets; florets closely imbricate, rachilla internodes 0.5–0.7 mm long; awns of the lowest lemmas 0.5–9 mm long .... *V. octoflora*
    2. Spikelets with 1–5 (–7) florets; florets not closely imbricate, rachilla internodes 1 mm or more long, visible between the florets; awns of lowest lemmas 4–20 mm long.
        3. Lowest panicle branches appressed-erect, without an axillary pulvinus .... *V. bromoides*
        3. Lowest panicle branches spreading or reflexed, with an axillary pulvinus .... *V. microstachys*

Vulpia bromoides **(L.) S.F. Gray** (*Festuca bromoides* L., *F. detonensis* [All.] Asch. & Graebn.)
Barren fescue
Habitat/Range: Mesic to dry rock outcrops, grassy bluffs, open forests, roadsides and disturbed areas in the lowland zone; common on S Vancouver Island, the Gulf Islands and the Queen Charlotte Islands, infrequent along the mainland coast; introduced from Europe.

Vulpia microstachys **(Nutt.) Munro ex Benth. var. *pauciflora* (Scribn. in Beal) Lonard and Gould** (*Festuca pacifica* Piper, *F. reflexa* Buckl.)
Small fescue
Habitat/Range: Dry meadows, waste places, fields and roadsides in the lowland, steppe and montane zones; frequent in SW BC, rare in SC BC; S to MT, UT, NM, CA and MX; S America.

Vulpia myuros **(L.) Gmel.** (*Festuca megalura* Nutt. = var. *hirsuta*, *F. myuros* L.)
Rattail fescue
Habitat/Range: Mesic to dry rock outcrops, grassy bluffs, open forests and disturbed areas in the lowland zone; common on S Vancouver Island and the Gulf Islands, infrequent on the Queen Charlotte Islands and lower mainland; var. *hirsuta* - S to AZ, NW CA and MX; S America, var. *myuros* - introduced from Eurasia and Africa.
Notes: Two varieties occur in BC.

1. Lemmas strongly ciliate at the apex .... var. *hirsuta* (Hack.) Asch. & Graebn.
1. Lemmas not ciliate .... var. *myuros*

Vulpia octoflora **(Walt.) Rydb.** (*Festuca octoflora* Walt. = var. *octoflora*; *F. octoflora* var. *glauca* [Nutt.] Fern. and var. *tenella* [Willd.] Fern. = var. *glauca*; *F. octoflora* var. *hirtella* [Piper] A.S. Hitchc. = var. *hirtella*)
Six-weeks fescue
Habitat/Range: Mesic to dry slopes, open forests and roadsides in the steppe and montane zones; frequent in SC and SE BC; E to PQ and S to NJ, FL, TX, MX and CA.
Notes: Three varieties occur in BC.

1. Spikelets typically 4–5.5 mm; awns of lowermost lemmas 0.3–3 mm .... var. *glauca* (Nutt.) Fern.
1. Spikelets typically 5.6–10 mm; awns of lowermost lemmas 2.5–9 mm.
    2. Lemmas glabrous or slightly scabrous, often scabrous on margins .... var. *octoflora*
    2. Lemmas long scabrous to pubescent .... var. *hirtella* (Piper) Henrard

43 Contributed by R.J. Lonard.

PONTEDERIACEAE

HETERANTHERA

***Heteranthera dubia* (Jacq.) MacMill.**
Water star-grass or water star-weed
Habitat/Range: Ponds, lakeshores and slow-moving streams in the steppe and montane zones; infrequent in SW and SC BC; E to PQ and S to FL, TX, CA and MX; S America.

POTAMOGETONACEAE[44]

POTAMOGETON

Species of the genus *Potamogeton* are variable and difficult to identify. For reliable identification try to collect specimens with fruits.

Stem anatomy provides a good guide for identification of broad-leaved pondweeds. Two anatomical characters were utilized in the key:

Interlacunar vascular bundles – vascular bundles within the aerenchyma of the stem. They can be absent (e.g., *P. alpinus, P. nodosus,* and *P. richardsonii*), present in one circle (e.g., *P. gramineus, P. oakesianus*), or are well developed throughout the aerenchyma (e.g., *P. amplifolius, P. natans, P. praelongus*).

Endodermis – a single layer of cells with thickened cell walls separating the central vascular bundles from aerenchyma. The cells of endodermis are either U-shaped (e.g., *P. gramineus*) or O-shaped (e.g., *P. illinoensis*).

To view these anatomical characters make a razor blade section of a stem from an area between the nodes and examine using a compound microscope under small magnification. Pieces of stem from dried herbarium specimens can be softened in hot water. For more information on anatomical characters see Ogden (1943) and Wiegleb (1990).

Achenes also provide a good guide for identification of pondweeds. Achene size, presence or absence of a beak at the tip of the achene and development of a keel at the dorsal side of the fruit are the main characters used in this key. For more on fruit morphology of pondweeds see Aalto (1970).

1. Stipules forming a sheath around the stem partly below the base of the leaf blade and partly above; the leaf attached 10 mm or more above the node.
    2. Leaves 3–4 mm broad, flat, rough on edges, stiff, with a broad midvein and more than 20 lateral veins .... *P. robbinsii*
    2. Leaves less than 3 mm broad, cylindrical, smooth, soft, with only one, hardly noticeable vein.
        3. Leaves acute and sharp pointed at apex; fruits 2.5–4 mm long with a short beak; the surface of fruits without radial striation (10x magnification) .... *P. pectinatus*
        3. Leaves obtuse and blunt at apex; fruits 2–3 mm long, beakless; the surface of fruits with radial striation (10x magnification).
            4. Primary stems 0.5–1 mm in diameter; all leaves filiform, 0.2–0.5 mm wide with tight sheaths; spikes with 2–5 whorls of flowers; fruits 2–2.5 mm long .... *P. filiformis*
            4. Primary stems 1–3 mm in diameter; lower leaves with blades 1–2 mm wide and with loose inflated sheaths, upper leaves from the branches filiform; spikes with 5–12 whorls of flowers; fruits 2.5–2.8 mm long .... *P. vaginatus*

---

44 Contributed by A. Ceska.

1. Stipules forming a sheath around the stem only above the base of the leaf blade; the leaf attached at the node.

    5. Submersed leaves linear, ribbon-like, or cylindrical, less than 5 mm wide, with parallel margins.

        6. Submersed leaves cylindrical, terete.

            7. Floating leaves 2.5–6 cm wide, usually cordate, larger ones with (18) 21–35 nerves; submersed leaves growing from the main stem; interlacunar vascular bundles in several circles throughout the aerenchyma; mature fruit (including beak) 3.7–4.5 mm long, obscurely keeled .... *P. natans*

            7. Floating leaves 1–3 cm wide, usually rounded or cuneate at base, largest ones with 11–19 nerves; submersed leaves growing from the branches of the main stem; interlacunar bundles developed only in the outer interlacunar circle of the aerenchyma; mature fruit (including beak) 3–3.4 mm long, prominently keeled .... *P. oakesianus*

        6. Submersed leaves flat.

            8. Submersed leaves with 9–35 veins .... *P. zosteriformis*

            8. Submersed leaves with 1–7 veins.

                9. Stipules strongly fibrous, becoming whitish, especially on the winter buds; the base of winter buds strongly ribbed.

                    10. Leaves obtuse or rounded and slightly mucronate, not conspicuously two-ranked; blades thin, 1.5–3.5 mm wide, flat, with 5–7 veins, with a narrow cellular-reticulate band along the midrib; winter buds fan-shaped; peduncles flattened .... *P. friesi*

                    10. Leaves gradually tapered into a sharp bristle tip, conspicuously two-ranked; blades firm, 0.5–2.5 mm wide, convolute, with 3 (rarely 5) veins, without any cellular-reticulate band along the midrib; winter buds slender; peduncles terete .... *P. strictifolius*

                9. Stipules delicate, not fibrous, greenish or brownish; the base of winter buds smooth.

                    11. Leaves 2–4 mm wide, rounded at apex; fruits 3–4 mm long ..... *P. obtusifolius*

                    11. Leaves 0.3–3 mm wide, acute to obtuse or mucronate; fruits 1.8–2.8 mm long.

                        12. Fruits with distinct dorsal keel; veins on stipules evident as ridges running all the length of the stipule; glands at the base of stipules either lacking or poorly developed .... *P. foliosus*

                        12. Fruits without dorsal keel; veins on stipules obscure and faint; glands at the base of stipules usually well developed.

                            13. Stipules connate, fused along the stem, at least when young; mature fruits broadest above the middle; plants sparsely branched .... *P. pusillus*

                            13. Stipules convolute; mature fruits broadest below the middle; plants branched .... *P. berchtoldii*

    5. Submersed leaves lanceolate or ovate, more than 5 mm wide.

        14. Leaf margins serrate; fruits with the beak as long as the body of the fruit or longer .... *P. crispus*

        14. Leaf margins entire; fruits with the beak much shorter than the body of the fruit.

15. Submersed leaves ribbon-like with parallel sides, 5–10 mm wide; parallel-sided median band of lacunae several cells wide and filling a fourth or more of the total blade width; stems compressed .... *P. epihydrus*

15. Submersed leaves lanceolate to ovate with sides ± curved most of their length; cellular-reticulate band absent or less than a fourth of the blade width; stems terete.

16. Submersed leaves sessile, cordate or rounded at the base and clasping the stems.

17. Leaves ovate-oblong, mostly 10–20 cm long, with cucullate apex; fruits more than 4 mm long; stems with many interlacunar vascular bundles .... *P. praelongus*

17. Leaves rounded, ovate, or elongate-ovate, 1–10 cm long, not with cucullate apex; fruits less than 3.5 mm long; stems lacking interlacunar vascular bundles.

18. Stipules coarse, disintegrating into persistent whitish fibers; peduncles clavate, 1.5–25 cm long; fruits with a cavity in the endocarp loop .... *P. richardsonii*

18. Stipules delicate, lacking on old specimens; peduncles not clavate, 1–9 cm long; fruits without a cavity in the endocarp loop *P. perfoliatus*

16. Submersed leaves petiolate or sessile, not clasping the stems.

19. Submersed leaves 2–5 cm broad, tapering into distinct petioles; stipules more than 3 cm long.

20. Floating leaves cordate; submersed leaves folded and strongly falcate, with more than 20 veins; interlacunar bundles strongly developed throughout .... *P. amplifolius*

20. Floating leaves rounded or cuneate at the base; submersed leaves flat and not falcate, with less than 20 veins; interlacunar bundles developed only in 1 circle or absent.

21. Petioles of submersed leaves shorter than 2 cm; interlacunar bundles well developed, forming one circle; endodermis of U-cells .... *P. illinoensis*

21. Petioles of submersed leaves longer than 2 cm; interlacunar bundles absent; endodermis of O-cells .... *P. nodosus*

19. Submersed leaves mostly less than 2 cm broad, sessile, or with petioles shorter than 0.5 cm; stipules less than 3 cm long.

22. Plants with reddish tinge, usually not branched; floating leaves, if present, not markedly different from the submersed leaves; interlacunar bundles absent; endodermis of O-cells .... *P. alpinus*

22. Plants greenish, freely branched; floating leaves markedly different from the submersed leaves; interlacunar bundles well developed in one circle; endodermis of U-cells .... *P. gramineus*

Potamogeton alpinus **Balbis** (*P. tenuifolius* Raf., *P. alpinus* var. *subellipticus* [Fern.] Ogden)
Northern pondweed
Habitat/Range: Nutrient poor lakes and sloughs in the lowland, steppe and montane zones; frequent throughout BC; circumpolar, N to AK, YT, NT, E to NF and S to NY, MI, WI, CO and CA, Eurasia.

Potamogeton amplifolius **Tuckerman**
Large-leaved pondweed
Habitat/Range: Lakes and ponds in the lowland and montane zones; frequent in SW BC, infrequent E of the Coast-Cascade Mountains S of 55°N; disjunct E from MB to NF and S to GA, AL, OK, MT, ID and CA.

Potamogeton berchtoldii **Fieb. in Bercht.** (*P. pusillus* auct. non L., *P. pusillus* var. *tenuissimus* Mert. & Koch, *P. berchtoldii* var. *lacunatus* [Haegstr.] Fern.)
Berchtold's pondweed
Habitat/Range: Lakes, ponds, sloughs and ditches in the lowland, steppe and montane zones; common throughout BC; circumpolar, N to AK, YT, NT, E to NF and S to GA, LA, TX and CA; Eurasia.

Potamogeton crispus **L.**
Curled pondweed
Habitat/Range: Nutrient rich lakes, ponds and sloughs in the lowland and steppe zones; common in S BC south of 50°N; introduced from Eurasia.

Potamogeton epihydrus **Raf.** (*P. nuttallii* Cham. & Schlecht., *P. epihydrus* var. *ramosus* [Peck] House)
Ribbon-leaved pondweed
Habitat/Range: Peaty lakes, ditches and ponds from the lowland and steppe to subalpine zones; common in coastal BC, infrequent E of the Coast-Cascade Mountains; N to AK, disjunct E from SK to NB and NS and S to VA, IN, IA, ID and CA; Europe (Outer Hebrides).

Potamogeton filiformis **Pers.**
Slender-leaved pondweed
Habitat/Range: Calcium rich lakes in the lowland, steppe and montane zones; frequent E of the Coast-Cascade Mountains, rare along the coast; circumpolar, N to AK, YT and NT, E to NF and S to ME, PA, MI, WI, SD, NV and CA; Eurasia, Australia.

Potamogeton foliosus **Raf.** (*P. foliosus* var. *macellus* Fern.)
Closed-leaved pondweed
Habitat/Range: Lakes, ponds and ditches in the lowland, steppe and montane zones; frequent in BC S of 55°N, rare in N BC; N to AK and YT, E to NF and S to SC, NM, CA and MX.

Potamogeton friesii **Rupr.** (*P. compressus* auct. non L.)
Flat-stalked pondweed
Habitat/Range: Lakes, ponds and ditches in the lowland, steppe and montane zones; frequent in BC east of the Coast-Cascade Mountains, infrequent in coastal BC; circumpolar, N to AK, YT and NT, E to NF and S to DE, LA, TX and CA, Eurasia.

Potamogeton gramineus **L.** (*P. gramineus* var. *maximus* Morong ex A. Benn, *P. gramineus* var. *myriophyllus* Robbins, *P. heterophyllus* Schreber)
Grass-leaved pondweed
Habitat/Range: Lakes, lake margins, ponds in peat bogs, ditches and slowly flowing streams and rivers in all but the alpine zone; frequent throughout BC; circumpolar, N to AK, YT and NT, E to NF and S to PA, MI, IA, NE, CO, NM, AZ and CA; Eurasia.
Notes: A highly variable species with several varieties described in N America based on stem branching and size of submersed leaves (var. *maximus* Morong ex A. Benn., var. *myriophyllus* Robbins). Their taxonomical value is questionable.

Potamogeton illinoensis **Morong**
Illinois pondweed
Habitat/Range: Lakes and slowly flowing streams in the steppe zone; infrequent in S BC east of Coast-Cascade Mountains; disjunct E from MA to PQ and S to VA (disjunct in FL), MO, TX, NM and CA.

***Potamogeton natans* L.**
Floating-leaved pondweed
Habitat/Range: Lakes and ponds from the lowland and steppe to subalpine zones; common throughout BC; cosmopolitan with disjunctions, N to AK, YT and NT, E to NF and S to NJ, PA, MI, IN, WI, NE, NM, UT, AZ and CA; Eurasia, S America, S Africa, Australia.

***Potamogeton nodosus* Poir. in Lam.**
Long-leaved pondweed
Habitat/Range: Lakes and sloughs of the lowland zone; infrequent in the lower Fraser River Valley, rare or doubtful elsewhere; cosmopolitan with disjunctions, E. disjunct to ON, PQ and NF and S to VA, AL, LA, TX and CA; Eurasia, Mexico, S America, Africa.

***Potamogeton oakesianus* Robbins**
Oakes' pondweed
Habitat/Range: Lakes in the lowland and steppe zones; rare in S BC (Steelhead, Mara Lake); disjunct E to ON, PQ and NF and S to NJ, MI and WI.

***Potamogeton obtusifolius* Mert. & Koch in Rohling**
Blunt-leaved pondweed
Habitat/Range: Lakes and sloughs in the lowland, steppe and montane zones; infrequent throughout BC; circumpolar, N to YT, E to ON, NS and NF and S to NJ, PA, MI and WI; Eurasia, Australia.

***Potamogeton pectinatus* L.**
Fennel-leaved pondweed
Habitat/Range: Lakes, ponds and estuaries in the lowland, steppe and montane zones; frequent in BC east of the Coast-Cascade Mountains, infrequent on the coast; cosmopolitan, N to AK, YT and NT, E to MB, ON, and NF and S to FL, TX and CA; Eurasia, Mexico, C and S Africa, C and S America, New Zealand, Australia.

***Potamogeton perfoliatus* L.**
Perfoliate pondweed
Habitat/Range: Lakes in the montane zone; rare in N BC (Cassiar Range - Swan Lake); circumpolar, disjunct E to PQ, NF and NB; Eurasia.

***Potamogeton praelongus* Wolfg.**
Long-stalked pondweed
Habitat/Range: Lakes in the lowland, steppe and montane zones; frequent throughout BC; circumpolar, N to AK and NT, E to NF and S to NJ, IN, NE, WY, CO and CA; Eurasia.

***Potamogeton pusillus* L.** (*P. panormitanus* Biv.-Bern., *P. pusillus* var. *minor* [Biv.] Fern. & Schub.)
Small pondweed
Habitat/Range: Lakes, ponds, sloughs and ditches in the lowland, steppe and montane zones; frequent in BC; circumpolar, E to PQ and S to AL, LA, TX and CA; Eurasia, Mexico, Africa.

***Potamogeton richardsonii* (A. Benn.) Rydb.** (*P. perfoliatus* var. *richardsonii* A. Benn.)
Richardson's pondweed
Habitat/Range: Lakes, ponds, sloughs in the lowland, steppe and montane zones; common throughout BC; amphiberingian, N to AK, YT and NT, E to NF and S to VT, PA, IN, NE, CO and CA; E Asia.

***Potamogeton robbinsii* Oakes**
Robbin's pondweed
Habitat/Range: Lakes in the lowland, steppe and montane zones; frequent in coastal BC, infrequent E of the Coast-Cascade Mountains; N to AK and NT, disjunct E to ON, PQ and NB and S to DE, IN, MN, WA and OR.

***Potamogeton strictifolius* A. Benn.**
Stiff-leaved pondweed
Habitat/Range: Lakes in the lowland and steppe zones; rare in SC and SE BC (Kawkawa Lake, Mara Lake and Windermere Lake); disjunct E to SK, ON and PQ and S to NY, IN, NE and MN.

***Potamogeton vaginatus* Turcz.**
Sheathing pondweed
Habitat/Range: Lakes in the montane zone; infrequent E of the Coast-Cascades Mountains; circumpolar, N to AK, YT and NT, E to SK, ON, and NF and S to NJ, MI, WI, ND, ID and OR; Eurasia.

***Potamogeton zosteriformis* Fern.**
Eel-grass pondweed
Habitat/Range: Lakes in the lowland, steppe and montane zones; frequent in BC east of the Coast-Cascade Mountains, infrequent in coastal BC; N to AK and NT, E to ON, PQ and NS and S to VA, IN, NE, MT, ID and CA.

RUPPIACEAE[45]

RUPPIA

***Ruppia maritima* L.** (*R. maritima* var. *longipes* Hagstr., var. *rostrata* Agardh and var. *spiralis* [L. ex Dumort.] Moris = var. *maritima*, *R. occidentalis* S. Wats. = var. *occidentalis*)
Ditch-grass
Habitat/Range: Tidal marshes, ponds, lakes and ditches in the lowland, steppe and montane zones; common along the coast and in SC BC, rare in SE BC; circumpolar, E to NF and S to FL and MX; Eurasia, S America (var. *maritima*); E to S MB and S to NE and MN (var. *occidentalis*).
Notes: This is a highly variable species, especially with respect to fruit characteristics. Treatments range from the recognition of only one variable species without infraspecific taxa (e.g., Hitchcock and Cronquist 1973, Dandy 1980) to the recognition of up to nine varieties (e.g., Scoggan 1978, Kartesz and Kartesz 1980, Brayshaw 1985). Two, sometimes intergrading, varieties are recognized for BC.

1. Stipular sheaths less than 2 cm long; fruits 2–3 mm long, usually asymmetric .... var. *maritima*
1. Stipular sheaths, or at least some of them, 2–7 mm long; fruits 3–4 mm long, symmetric ...... var. *occidentalis* (Wats.) Graebn.

SCHEUCHZERIACEAE

SCHEUCHZERIA

***Scheuchzeria palustris* L. var. *americana* Fern.**
Scheuchzeria
Habitat/Range: Lakeshores and bogs in the lowland and montane zones; frequent in SW, C and E BC, also along the N Coast; N to AK, YT and NT, E to NF and S to PA, OH, NE, NM and CA.

45 Sometimes included in Zosteraceae.

SPARGANIACEAE

SPARGANIUM

1. Stigmas 2, rarely 1, more than 2 mm long; achenes broadly obpyramidal, somewhat truncate at the top; staminate heads occurring above the pistillate heads .... *S. eurycarpum*

1. Stigmas 1, less than 2 mm long; achenes ellipsoid, tapering at each end; staminate heads occurring below the pistillate heads.

    2. Achene beaks less than 1.5 mm long or beakless; staminate head solitary.

        3. Achene beaks less than 0.5 mm long or beakless; staminate head closely adjacent to the upper pistillate head; basal leaves (0.5) 1–3 (5) mm wide, opaque, often yellow .... *S. hyperboreum*

        3. Achene beaks 0.5–1.5 mm long; staminate head distant from the upper pistillate head; basal leaves (1.5) 2–6 (10) mm wide, translucent, usually dark green .... *S. natans*

    2. Achene beaks 1.5–5 mm long; staminate heads solitary to several.

        4. Achene beaks conspicuously curved .... *S. fluctuans*

        4. Achene beaks straight or slightly curved.

            5. Staminate head usually solitary, closely adjacent to the upper pistillate head .... *S. glomeratum*

            5. Staminate heads (1) 3–8, distant from the upper pistillate head.

                6. Pistillate heads (1) 2–3 (4), usually crowded and appearing as one elongated head terminating the inflorescence, beaks 1.5–2.0 (2.2) mm long .... *S. angustifolium*

                6. Pistillate heads (3) 4–7 (10), distant and distinct; beaks 2–4.5 (6) mm long .... *S. emersum*

Sparganium angustifolium **Michx.** (*S. angustifolium* ssp. *emersum* [Rehm.] T.C. Brayshaw, var. *multipedunculatum* [Morong] T.C. Brayshaw; *S. emersum* var. *angustifolium* [Michx.] Taylor & MacBryde, *S. multipedunculatum* (Morong) Rydb., *S. simplex* var. *multipedunculatum* Morong)

Narrow-leaved bur-reed

Habitat/Range: Ponds, ditches and lakeshores in all but the alpine zone; common throughout BC; circumpolar, N to AK, YT an NT, E to NF and S to NC, IA, MN, CO and CA; Greenland, Europe.

Notes: This is a highly variable species in which different forms are often treated as species (e.g., Hitchcock and Cronquist 1973) or infraspecific taxa (e.g., Brayshaw 1985). Although Packer (1983b) presented evidence to show that there is no justification for recognizing more than a single, variable species, recent research by Cook and Nicholls (1986) has shown that *S. emersum* is best kept separate.

Sparganium emersum **Rehm. ssp. *emersum*** (*S. angustifolium* ssp. *emersum* [Rehm.] T.C. Brayshaw, *S. angustifolium* ssp. *emersum* var. *chlorocarpum* [Rydb.] T.C. Brayshaw, *S. chlorocarpum* Rydb., *S. simplex* Huds.)

Emersed bur-reed

Habitat/Range: Streams, ditches, ponds and lakeshores in the lowland, steppe and montane zones; common in coastal BC, less frequent elsewhere; circumpolar, N to AK, YT and NT, E to NB and NS and S to MI, WY, CO, NM and CA; Eurasia.

Sparganium eurycarpum **Engelm. in A. Gray** (*S. eurycarpum* var. *greenei* [Morong] Graebn.)

Broad-fruited bur-reed

Habitat/Range: Wet meadows, shallow ponds and lakeshores in the lowland, steppe and montane zones; infrequent in S and E BC; N to NT, E to NF and S to FL, CA and MX.

Sparganium fluctuans **(Morong) B.L. Robins.**
Water bur-reed
Habitat/Range: Ponds, lakeshores and slow-moving streams in the lowland and montane zones; rare in SW and C BC; circumpolar, E disjunctly to NF and S to NJ, TN and UT; Eurasia.

Sparganium glomeratum **Laest. ex Beurl.**
Glomerate bur-reed
Habitat/Range: Ponds and lakeshores in the lowland and montane zones; rare in WC BC, known only from Buck Channel, Queen Charlotte Islands and the Smithers area; E disjunctly to PQ and S to MN.

Sparganium hyperboreum **Laest. ex Beurl.**
Northern bur-reed
Habitat/Range: Ponds and lakeshores in the lowland and montane zones; infrequent throughout BC; circumpolar, N to AK, YT and NT and E to NF; Greenland, Eurasia.

Sparganium natans **L.** (*S. minimum* [Hartm.] Wallr.)
Small bur-reed
Habitat/Range: Ponds, lakeshores and slow-moving streams in the lowland and montane zones; common throughout all but NW BC; circumpolar, N to AK, YT and NT, E to NF and S to NJ, PA, MN, TN, UT and OR; Eurasia.

TYPHACEAE

TYPHA

1. Staminate and pistillate portion of spikes separate; pollen shed singly .... *T. angustifolia*
1. Staminate and pistillate portion of spikes usually continuous; pollen shed in tetrads .... *T. latifolia*

Typha angustifolia **L.**
Lesser cattail
Habitat/Range: Wet sites in the lowland zone; rare in the Vancouver area; introduced from E North America.

Typha latifolia **L.**
Common cattail or common bulrush
Habitat/Range: Wet sites, ponds, lakeshores and marshes in the lowland, steppe and montane zones; common in S BC, less frequent in NE BC; circumpolar, N to AK, YT an NT, E to NF and S to FL, LA, TX, NM, CA and MX; Eurasia, N Africa.

ZANNICHELLIACEAE[46]

ZANNICHELLIA

Zannichellia palustris **L.**
Horned pondweed
Habitat/Range: Shallow, often calcareous or brackish lakeshores, slow-moving streams and tidal marshes in the lowland, montane and steppe zones; frequent in mainland BC south of 54°N, less frequent northward and on Vancouver Island; cosmopolitan, E to NF and NS and S to FL, TN, TX, MX; S America, Eurasia, Africa.

---

46 Sometimes included in Zosteraceae.

ZOSTERACEAE

1. Plants dioecious; spadix bordered by conspicuous flap-like projections; rhizomes short, stout .... *Phyllospadix*

1. Plants monoecious; spadix with inconspicuous projections, if any; rhizomes elongate, slender .... *Zostera*

PHYLLOSPADIX[47]

1. Stems of fertile shoots 4–12 dm long, branched, with spathes usually paired at the nodes ... *P. torreyi*
1. Stems of fertile shoots 0.5–4 dm long, unbranched, with spathes usually solitary at the nodes.
    2. Leaves with 3, rarely 5, longitudinal veins, margins entire .... *P. scouleri*
    2. Leaves with 5, sometimes 7, longitudinal veins, margins toothed toward the apex .... *P. serrulatus*

***Phyllospadix scouleri* Hook.**
Scouler's surf-grass
Habitat/Range: Exposed, rocky, intertidal to subtidal coastal shores; common along the coast; N to AK and S to MX.

***Phyllospadix serrulatus* Rupr. ex Aschers.**
Toothed surf-grass
Habitat/Range: Sheltered intertidal and subtidal coastal shorelines; frequent along the coast; N to AK and S to OR.
Notes: Until recently (Phillips 1970), material of this species was included in *P. scouleri*.

***Phyllospadix torreyi* S. Wats.**
Torrey's surf-grass
Habitat/Range: Exposed, rocky, intertidal to subtidal coastal shores; infrequent in coastal BC; S to MX.

ZOSTERA[48]

1. Leaves 1–1.5 mm wide, 3-veined; leaf sheaths all split to the base .... *Z. japonica*
1. Leaves 1.5–12 mm wide, 5-many veined; leaf sheaths on sterile shoots closed at the base .... *Z. marina*

***Zostera japonica* Aschers. & Graebn.** (*Z. americana* den Hartog, *Z. nana* auct. non [Mertens] Roth, *Z. noltii* auct. non Hornem.)
Japanese eel-grass
Habitat/Range: Tidal mudflats along the coast; rare in SW BC, known only from Boundary Bay and Tsawwassen; introduced from E Asia.
Notes: Several recent papers have shown that *Z. americana* (recently named by den Hartog [1970]) is not readily separable from either *Z. noltii* (*Z. nana*) (Phillips and Shaw 1976) or *Z. japonica* (Harrison 1976, Bigley and Barreca 1982). Until adequate research has been conducted on this complex, the hypothesis of Harrison (1976), that our species (*Z. japonica*) was likely introduced from E. Asia, appears more plausible.

---

47 Key adapted from Brayshaw (1985).
48 Key adapted from Brayshaw (1985).

Zostera marina **L.** (*Z. marina* var. *latifolia* Morong and var. *stenophylla* [Raf.] Aschers. & Graebn., *Z. oregana* S. Wats., *Z. pacifica* S. Wats., *Z. stenophylla* Raf.)

Common eel-grass

Habitat/Range: Sheltered coastal waters; common in coastal BC; cosmopolitan, N to AK, S to CA; E North America, Greenland, Europe, E Asia, N Africa.

Notes: Recognition of infraspecific taxa in *Z. marina*, based on leaf features, appears to be a meaningless exercise considering the plasticity of these features and the lack of research.

REFERENCES

GENERAL

Abrams, L.R. 1923–1951. Illustrated flora of the Pacific States: Washington, Oregon and California. Vols. 1–3. Stanford Univ. Press, Stanford.

Argus, G.W. 1973. The genus *Salix* in Alaska and the Yukon. Nat. Mus. Natur. Sci. Publ. Bot., No. 2; Nat. Mus. Can., Ottawa. 279 p.

Bailey, L.H. 1949. Manual of cultivated plants. The Macmillan Co., New York.

Boivin, B. 1966–1967. Enumeration des plants du Canada. Provancheria No. 6. Univ. Laval, Quebec.

Boivin, B. 1968–1969. Flora of the Prairie Provinces. Provancheria No. 3. Univ. Laval, Quebec.

Brayshaw, T.C. 1976. Catkin bearing plants of British Columbia. B.C. Prov. Mus. Occas. Pap. Ser. No. 18. 176 p.

Brayshaw, T.C. 1985. Pondweeds and bur-reeds, and their relatives: Aquatic families of Moncotyledons in British Columbia. B.C. Prov. Mus. Occas. Pap. Series No. 26. Victoria.

Brayshaw, T.C. 1989. Buttercups, waterlilies, and their relatives: (The Order Ranales) in British Columbia. Royal B.C. Mus. Mem. No. 1. 253 p.

Calder, J.A. and R.L. Taylor. 1968. Flora of the Queen Charlotte Islands, Part 1. Can. Dept. Agric. Res. Branch Monogr. 4. 659 p.

Campbell, R. 1904. Some conspicuous British Columbia plants. Can. Rec. Sci. 9:176–189.

Clark, L.J. 1973. Wild flowers of British Columbia. Gray's Publ. Ltd., Sidney, British Columbia. 591 p.

Clark, L.J. 1976. Wild flowers of the Pacific Northwest. Gray's Publ. Ltd., Sidney, British Columbia. 604 p.

Cody, W.J. and D.M. Britton. 1989. Ferns and fern allies of Canada. Agri. Can. Res. Br., Publ. 1829/E. Can. Gov. Publ. Centre, Ottawa, Ont. 430 p.

Correll, D.S. and H.B. Correll. 1975. Aquatic and wetland plants of southwestern United States. Vols. 1 and 2. Stanford Univ. Press, Stanford.

Dansereau, P. 1957. Biogeography. Ronald Press, New York.

Daubenmire, R. 1970. Steppe vegetation of Washington. Wash. Agric. Exp. Stn. Tech. Bull. 62.

Davis, R.J. 1952. Flora of Idaho. Wm. C. Brown Co., Dubuque, Iowa. 828 p.

Douglas, G.W. 1971. The alpine–subalpine flora of the North Cascade Range, Washington. Wasmann J. Biol. 29:129–168.

Douglas, G.W. 1972. Subalpine plant communities of the western North Cascades, Washington. Arct. Alp. Res. 4:147–166.

Douglas, G.W. 1974. Montane zone vegetation of the Alsek River region, southwestern Yukon. Can. J. Bot. 52:2505–2532.

Douglas, G.W. 1982. The sunflower family (Asteraceae) of British Columbia. Vol. I. Senecioneae. B.C. Prov. Mus. Occas. Pap. Ser. No. 23.

Douglas, G.W. 1993. The sunflower family (Asteraceae) of British Columbia. Vol. II – Astereae, Anthemideae, Eupatoreae and Inuleae. B.C. Prov. Mus. Occas. Pap. Ser. (in press).

Douglas, G.W. 1994. The sunflower family (Asteraceae) British Columbia. Vol. III. Lactuceae, Cardueae, and Heliantheae. B.C. Prov. Mus. Occas. Pap. Ser. (in press).

Douglas, G.W., G.W. Argus, H.L. Dickson and D.F. Brunton. 1981. The Rare Vascular Plants of the Yukon. Syllogeus No. 28.

Douglas, G.W. and J.G. Packer. 1988. *Erigeron salishii*, a new *Erigeron* (Asteraceae) from British Columbia and Washington. Can. J. Bot. 66:414–416.

Douglas, G.W. and L.C. Bliss. 1977. Alpine and high subalpine plant communities of the North Cascades Range, Washington and British Columbia. Ecol. Monogr. 47: 113–150.

Douglas, G.W., A. Ceska and G.G. Ruyle. 1983. A floristic bibliography for British Columbia. Land Manag. Rep. No. 15. Victoria, B.C. Min. For.

Eastham, J.W. 1947. Supplement to the Flora of Southern British Columbia B.C. Prov. Mus. Spec. Publ. No. 1, Victoria. 119 p.

Fernald, M.L. 1950. Gray's manual of Botany. 8th ed. American Book Co., New York, N.Y. 1632 p.

Ferris, R.S. 1960. Illustrated Flora of the Pacific States: Washington, Oregon and California. Vol. IV. Bignoniaceae to Compositae. Stanford Univ. Press, Stanford.

Franklin, F.J. and C.T. Dyrness. 1973. Natural vegetation of Oregon and Washington. U.S.D.A. For. Serv. Gen. Tech. Rep. PNW–8.

Frankton, C. and G.A. Mulligan. 1970. Weeds of Canada. Rev. Ed., Can. Dept. Agric. Publ. 948. Queen's Printer, Ottawa. 217 p.

Gray, A. 1884. Caprifoliaceae to Compositae. *In* Synoptical flora of North America. Vol. 1. Amer. Book Co., New York.

Groh, H. and C. Frankton. 1949. Canadian weed survey. Sixth Annual Report. Can. Dept. Agric., Ottawa.

Henderson, D.M., R.K. Moseley and A.F. Cholewa. 1990. A new *Agoseris* (Asteraceae) from Idaho and Montana. System. Bot. 15:462–465.

Henry, J.K. 1915. Flora of Southern British Columbia. W.J. Gage and Co. Ltd., Toronto. 363 p.

Hitchcock, C.L. and A. Cronquist. 1973. Flora of the Pacific Northwest. Univ. Wash. Press, Seattle.

Hitchcock, C.L., A. Cronquist, M. Ownbey and J.W. Thomson. 1955–1969. Vascular plants of the Pacific Northwest. Parts 1–5. Univ. Wash. Press, Seattle.

Hosie, R.C. 1969. Native trees of Canada. 7th ed. Can. For. Serv., Dept. Environ. Queen's Printer, Ottawa. 380 p.

Hulten, E. 1937. Outlineof the history of arctic and boreal biota during the Quaternary Period. Bokförlags Aktiebolaget, Thule.

Hulten, E. 1968. Flora of Alaska and neighboring territories. Stanford Univ. Press, Stanford. 1008 p.

Kartesz, J.T. and R. Kartesz. 1980. A synonymized checklist of the vascular flora of the United States, Canada, and Greenland. Vol. II. The biota of North America. Univ. North Carolina Press, Chapel Hill, N.C. 498 p.

Krajina, V. 1965. Biogeoclimatic zones and classification of British Columbia. Ecol. West. N. Amer. 1:1–17.

Krajina, V. 1969. Ecology of forest trees in British Columbia. Ecol. West. N. Amer. 2:1–146.

Lellinger, D.B. 1985. A field manual of the ferns & fern-allies of the United States & Canada. Smithsonian Inst. Press, Washington, D.C. 389 p.

Lõve, D. 1970. Subartic and subalpine: where and what? Arct. Alp. Res.2:63–73.

Macoun, J. 1883–1890. Catalogue of Canadian plants. Parts 1–5. Geol. Surv. Can., Ottawa.

Macoun, J.M. 1889. Checklist of Canadian Plants. Cunningham and Lindsay, Ottawa.

Macoun, J.M. 1894–1906. Contributions to Canadian botany. Parts 1–18. Ottawa Field-Natur., Ottawa.

Meusel, H., E. Jager and E. Weinert. 1965. Vergleichende chorologie der zentraleuropaischen flora. I. Gustav. Fischer, Jena.

Packer, J.G. 1983. Flora of Alberta. Second Edition. Univ. Toronto Press, Toronto, Ont.

Page, C.N. 1982. The ferns of Britain and Ireland. Cambridge University Press, Cambridge. 447 p.

Porsild, A.E. 1955. The vascular plants of the western Canadian Arctic Archipelago. Bull. Natl. Mus. Canada 135.

Porsild, A.E. and W.J. Cody. 1980. Vascular plants of continental Northwest Territories, Canada. Nat. Mus. Can., Ottawa. 667 p.

Raven, P.H. 1963. Amphitropical relationships in the floras of North and South America. Quart. Rev. Biol. 38:151–177.

Rowe, J.S. 1959. Forest regions of Canada. Can. Dept. North. Affairs Nat. Resour. For. Branch Bull. 123.

Rowe, J.S. 1972. Forest regions of Canada. Can. For. Serv. Publ. 1300.

Rydberg, P.A. 1922. Flora of the Rocky Mountains and adjacent plains. 2nd ed. P.A. Rydberg, New York. 1143 p.

Savale, W.F., Jr. and A. Ceska. 1989. Noteworthy collections, British Columbia: *Wolffia columbiana* Karsten (Lemanaceae). Madrono 36:153.

Schofield, W.B. 1969. Phytogeography of Northwestern North America: Brophytes and Vascular Plants. Madrono 209:155–207.

Scoggan, H.J. 1978–1979. The flora of Canada. Parts 1–4. Nat. Mus. Nat. Sci. Publ. Bot., No. 7:1–1626.

Steward, A.N., L.J. Dennis and H.M. Gilkey. 1963. Aquatic plants of the Pacific Northwest. Oregon State Univ. Press, Corvallis.

Straley, G.B., R.L. Taylor and G.W. Douglas. 1985. The rare vascular plants of British Columbia. Nat. Mus. Can., Syllogeus 59, Ottawa.

Szczawinski, A.F. 1959. The orchids of British Columbia. B.C. Prov. Mus. Handb. 16, Victoria. 124 p.

Szczawinski, A.F. 1962. The heather family (Ericaceae) of British Columbia. B.C. Prov. Mus. Handb. 10, Victoria. 205 p.

Szczawinski, A.F and A.S. Harrison. 1973. Flora of Saanich Peninsula. B.C. Prov. Mus. Occas. Pap. Ser. No. 16, Victoria.

Taylor, R.L. and B. MacBryde. 1977. Vascular plants of British Columbia. Bot. Gard. Tech. Bull. No. 4. Univ. of B.C. Press, Vancouver. 754 p.

Taylor, T.M.C. 1963. The ferns and fern allies of British Columbia. B.C. Prov. Mus. Handb. 12. Victoria. 172 p.

Taylor, T.M.C. 1966. The lily family (Liliaceae) of British Columbia. B.C. Prov. Mus. Handb. 25, Victoria. 109 p.

Taylor, T.M.C. 1970. Pacific Northwest ferns and their allies. Univ. of Toronto Press, Toronto. 247 p.

Taylor, T.M.C. 1973. The rose family (Rosaceae) of British Columbia. B.C. Prov. Mus. Handb. 30. Victoria. 223 p.

Taylor, T.M.C. 1974a. The pea family (Leguminosae) of British Columbia. B.C. Prov. Mus. Handb. 32, Victoria. 251 p.

Taylor, T.M.C. 1974b. The figwort family (Scrophulariaceae) of British Columbia. B.C. Prov. Mus. Handb. 33, Victoria. 237 p.

Taylor, T.M.C. 1983. The sedge family (Cyperaceae). B.C. Prov. Mus. Handb. 43, Victoria. 375 p.

Tutin, T.G., V.H. Heywood, N.A. Burges, D.M. Moore, D.H. Valentine, S.M. Walters and D.A. Webb. [eds.]. 1964–1980. Flora Europaea. Volumes 1–5. Univ. Cambridge Press, Cambridge.

Ulke, T.A. 1935. List of the vascular plants of the Horsethief Creek – Purcell Range, B.C. Can. Field-Natur. 49:49–55, 71–76.

Wallace, G.D. 1975. Studies of the Monotropoideae (Ericaceae): taxonomy and distribution. Wasmann J. Biol. 33:1–88.

Welsh, S.L. 1974. Anderson's flora of Alaska and adjacent parts of Canada. Brigham Young Univ. Press, Provo. 724 p.

ALISMATACEAE

Bjorkqvist, I. 1967. Studies in *Alisma* L. I. Distribution, variation and germination. Opera Bot. 17:1–128.

Bjorkqvist, I. 1968. Studies in *Alisma* L. II. Chromosome studies, crossing experiments and taxonomy. Opera Bot. 19:1–138.

Bogin, C. 1955. A revision of the genus *Sagittaria* (Alismataceae). Mem. New York Bot. Gard. 9:179–233.

Fernald, M.L. 1946. The North American representative of *Alisma plantago-aquatica*. Rhodora 48:86–88.

Frey, L. 1966. Embryological studies on *Alisma lanceolatum* With. Act. Biol. Crac. 9:125–136.

Hendricks, A.J. 1957. A revision of the genus *Alisma* (Dill.) L. Amer. Midl. Nat. 58:470–493.

Pogan, E. 1963. Taxonomic value of *Alisma triviale* Pursh. and *Alisma subcordatum* Rafin. Can. J. Bot. 41:1011–1013.

Reveal, J.L. 1977. Alismataceae. *In* A. Cronquist *et al.* Vascular plants of the Intermountain West, U.S.A. Columbia Univ. Press, New York.

Rubtzoff, P. 1964. Notes on the genus *Alisma*. Leafl. West. Bot. 10:90–95.

Samuelson, G. 1932. Die Arten der Gattung *Alisma* L. Ark. Bot. 24A:1–46.

Wooten, J.W. 1973. Taxonomy of seven species of *Sagittaria* from eastern North America. Brittonia 25:64–74.

ARACEAE

Hulten, E. and H. St. John. 1931. The American species of *Lysichitum*. Sv. Bot. Tidskr. 25:453–464.

Packer, J.G. and G.S. Ringius. 1984. The distribution and status of *Acorus* (Araceae) in Canada. Can. J. Bot. 62:2248–2252.

CYPERACEAE

Arwidsson, T. 1928. Bemerkungen zur Androgynie in der *Carex dioica* Gruppe. Svensk Bot. Tidskr. 22:106–114.

Beetle, A.A. 1940. Studies in the genus *Scirpus* L.: I. Delimination of the subgenera Euscirpus and Aphylloides. Amer. J. Bot. 27:63–64.

Beetle, A.A. 1941. Studies in the genus *Scirpus* L.: II. The section Baeothryon Ehrh. Amer. J. Bot. 28:469–476.

Beetle, A.A. 1941. Studies in the genus *Scirpus* L.: III. The American species of the section Lacustres Clarke. Amer. J. Bot. 28:691–700.

Beetle, A.A. 1942. Studies in the genus *Scirpus* L.: IV. The section Bolboschoenus Palla. Amer. J. Bot. 29:82–88.

Beetle, A.A. 1942. Studies in the genus *Scirpus* L.: V. Notes on the section Actaeogeton Reich. Amer. J. Bot. 29:653–656.

Beetle, A.A. 1943. A key to the North American species of the genus *Scirpus* based on achene characters. Amer. Midl. Nat. 29:533–538.

Beetle, A.A. 1943. Studies in the genus *Scirpus* L.: VI. The section Schoenoplectus Palla. Amer. J. Bot. 30:395–401.

Beetle, A.A. 1944. Studies in the genus *Scirpus* L.: VII. Conspectus of sections represented in the Americas. Amer. J. Bot. 31:261–265.

Beetle, A.A. 1946. Studies in the genus *Scirpus* L.: VIII. Notes on its taxonomy, phylogeny, and distribution. Amer. J. Bot. 33:660–666.

Beetle, A.A. 1947. *Scirpus* L. *In* North American Flora 18:481–504.

Bergren, G. 1961. Atlas of seeds and small fruits of northwestern European plant species with morphological descriptions. II. Cyperaceae. Swedish Natural Science Research Council, Stockholm.

Boott, F. 1858–1897. Illustrations of the genus *Carex*. Part I–IV. William Pamplin, Soho. Reprint 1968, J. Cramer Verlag, Lehre.

Böcher, T.W. 1950. Contributions to the flora and plant geography of west Greenland. II. The *Carex capitata*-, the *Luzula multiflora*-, and the *Tortularia humilis*-complexes. Meddel. Gronland Bd. 147:1–39.

Böcher, T.W. 1952. A study of the circumpolar *Carex heleonastes-amblyorhyncha* complex. Acta Arctica 5:1–32.

Calder, J.A. 1952. Notes on the genus *Carex*. I: A new species of *Carex* from western Canada. Rhodora 54:246–250.

Chase, A. 1904. The North American allies of *Scirpus lacustris*. Rhodora 6:65–71.

Crins, W.J. and P.W. Ball. 1983. The taxonomy of the *Carex pensylvanica* complex (Cyperaceae) in North America. Can. J. Bot. 61:1692–1717.

Crins, W.J. and P.W. Ball. 1989. Taxonomy of the *Carex flava* complex (Cyperaceae) in North America and northern Eurasia. I. Numberical taxonomy and character analysis. Can. J. Bot. 67:1032–1047.

Crins, W.J. and P.W. Ball. 1989. Taxonomy of the *Carex flava* complex (Cyperaceae) in North America and northern Eurasia. II. Taxonomic treatment. Can. J. Bot. 67:1048–1065.

Davis, E.W. 1956. Cytology, evolution and origin of the aneuploid series in the genus *Carex*. Hereditas 42:349–365.

Donovan, J.W. 1976. *Eleocharis palustris*. Bot. Soc. Brit. Isles News 14:18–19.

Drapalik, D.J. and R.H. Mohlenbrock. 1960. A study of *Eleocharis*, Series Ovatae. Amer. Midl. Naturalist 64:339–341.

Duman, M.G. 1941. The genus *Carex* in eastern arctic Canada. Contr. Biol. Lab. Catholic Univ. Amer. 36:1–84.

Drury, W.H. Jr. 1956. The ecology of the natural origin of a species of *Carex* by hybridization. Rhodora 58:51–72.

Eastham, J.W. n.d. Key to British Columbia species of *Carex*. Unpublished mimeo. Univ. of British Columbia. 9 p.

Egorova, T.V. 1966. Osoki SSSR. Vidy podroda *Vignea*. Moskva & Leningrad. 266 p.

Faulkner, J.S. 1973. Experimental hybridization of northwestern European species in *Carex* sect. Acutae (Cyperaceae). J. Linn. Soc. Bot. 67:233–252.

Fernald, M.L. 1906. The varieties of *Carex paupercula*. Rhodora 8:73–77.

Fernald, M.L. 1929. The representatives of *Eleocharis palustris* in North America. Rhodora 31:57–78.

Fernald, M.L. 1942. Critical notes on *Carex*. Rhodora 44:281–330.

Ford, B.A. and P.W. Ball. 1992. The taxonomy of the circumpolar short-beaked taxa of Carex sect. *Vesicariae* (Cyperaceae). Syst. Bot. 17:620–639.

Gubanov, I.A., J.E. Alexeev, M.G. Vakhremeeva and V.S. Novikov (eds.). 1980. Rod Osoka. *In* Rabotnov, T.A. (ed.) Biologicheskaya flora Moskovskoy oblasti, vyp. 6:1–221. Izd. Moskovskogo Univ., Moskva.

Hakansson, A. 1954. Meiosis and pollen mitosis in X-rayed and untreated spikelets of *Eleocharis palustris*. Hereditas 40:325–345.

Hakansson, A. 1958. Holocentric chromosomes in *Eleocharis*. Hereditas 44:531–540.

Halliday, G. and A.O. Chater. 1969. Studies in the *Carex glareosa* complex. I. Fruit-shape. Feddes Repert. 80:77–91.

Halliday, G. and A.O. Chater. 1969. *Carex marina* Dewey, an earlier name for *C. amblyorhyncha* Krecz. Feddes Repert. 80:93–101.

Hermann, F.J. 1940. The genus *Carex* in Michigan. Amer. Midl. Naturalist 25:1–72.

Hermann, F.J. 1954. Addenda to North American Carices. Amer. Midl. Naturalist 58:470–493.

Hermann, F.J. 1957. New carices from the Canadian Rocky Mountains. Leafl. West. Bot. 8:109–114.

Hermann, F.J. 1968. Notes on Rocky Mountain Carices. Rhodora 70:419–421.

Hermann, F.J. 1970. Manual of the Carices of the Rocky Mountains and Colorado Basin. USDA Agric. Handbook No. 374. 397 p.

Hermann, F.J. 1974. Manual of the genus *Carex* in Mexico and Central America. USDA Agric. Handbook No. 467. 219 p.

Holm, T. 1900. Studies in the Cyperaceae, XIV. Amer. J. Sci. 10:266–284.

Hudson, J.H. 1977. *Carex* in Saskatchewan. Bison Publishing House, Saskatoon, Sask. 193 p.

Jermy, A.C. and T.G. Tutin. 1968. British sedges. A handbook to the species of *Carex* found growing in the British Isles. Bot. Soc. Brit. Isles, London. 199 p.

Kalela, A. 1944. Systematische und pflanzengeographische Studien an der *Carex*-Subsektion Alpinae Kalela. Ann. Bot. Soc. Zool. Bot. Fenn. Vanamo 19:1–128.

Kalela, A. 1950. *Carex media* R.Br. der älteste Name einer zirkumpolar-borealen Seggenart. Arch. Soc. Zool. Bot. Fenn. Vanamo 4:114–115.

Kalela, A. 1965. Über die Kollektivart *Carex brunescens* (Pers.) Poir. Ann. Bot. Fenn. 2:174–218.

Kelso, L. 1953. The *Carex* species of Colorado. Biol. Leafl. 66. 38 p.

Koyama, T. 1961. Taxonomic studies of Cyperaceae. 12. Naturaliste Can. 88:237–252.

Koyama, T. 1962. The genus *Scirpus* Linn.: Some North American aphylloid species. Can. J. Bot. 40:913–937.

Koyama, T. 1963. The genus *Scirpus* Linn.: Critical species of the section Pterolepis. Can. J. Bot. 41:1107–1131.

Koyama, T. 1966. Classification of the family Cyperaceae (3). Quart. J. Taiwan Mus. 14:159–194.

Koyama, T. 1980. The genus *Bolboschoenus* Palla in Japan. Acta Phytotax. Geobot. 31:139–148.

Krechetovich, V.I. 1935. Genus *Carex* L. *In* Komarov, V.L. (ed.) Flora of the USSR, III:111–464, 589–623. Leningrad. [English translation III:86–369. Jerusalem 1964].

Krechetovich, V.I. 1936. Are the sedges of the subgenus Primocarex Kuk. primitive? Bot. Zhurnal 21:395–424.

Kükenthal, G. 1909. Cyperaceae – Caricoidae. *In* Engler, A. (ed.): Das Pflanzenreich 4: (20,H 38). Leipzig. 824 p.

Lewis, K.R and B. John. 1961. Hybridization in a wild population of *Eleocharis palustris*. Chromosoma 12:433–448.

Löve, A., D. Löve and M. Raymond. 1957. Cytotaxonomy of *Carex* section Capillares. Can. J. Bot. 35:715–761.

Mackenzie, K.K. 1931–1935. Cariceae. *In* North American Flora 18:1–478.

Mackenzie, K.K. 1940. North American Cariceae, Vol. I and II. New York Botanical Garden.

Mackenzie, K.K. 1960. Keys to the North American species of *Carex,* North American flora, Vol. 18 parts 1–7, reprinted for the New York Botanical Garden by Stechert-Hafner Service Agency, Inc.

McGivney, Sister M.V.D.P. 1938. A revision of the subgenus Eucyperus found in the United States. Catholic Univ. Amer. Biol. Ser. No. 26. 74 p.

Metcalfe, C.R. 1971. Anatomy of the Monocotyledons, V. Cyperaceae. Oxford Univ. Press, London. 597 p.

Moore, D.M. and A.O. Chater. 1971. Studies of bipolar disjunct species. I. *Carex.* Bot. Not. 124:317–334.

Murray, D.F. 1969. Taxonomy of *Carex*, sect. Atratae (Cyperaceae) in the southern Rocky Mountains. Brittonia 21:55–76.

Murray, D.F. 1970. *Carex podocarpa* and its allies in North America. Can. J. Bot. 48:313–324.

Nilsson, O. and H. Hjelmqvist. 1967. Studies on the nutlet structure of southern Scandinavian species of *Carex.* Bot. Not. 120:460–485.

Norlindh, T. 1972. Notes on the variation in the *Scirpus maritimus* complex. Bot. Not. 125:397–405.

Raymond, M. 1949. Notes sur le genre *Carex*: II. La valeur taxonomique de *C. arctogena*. Contr. Inst. Bot. Univ. Montreal 64:37–41.

Raymond, M. 1951. Sedges as material for phytogeographic studies. Mem. Jard. Bot. Montreal 20:1–23.

Raymond, M. 1952. The identity of *Carex misandroides* Fern. with notes on the North American Frigidae. Can. Field-Naturalist 66:95–103.

Reznicek, A.A. and P.W. Ball. 1981. The sedge *Carex loliacea* in eastern North America. Can. Field-Naturalist 95:89–92.

Reznicek, A.A. and P.W. Ball. 1980. The taxonomy of *Carex* section Stellulatae in North America north of Mexico. Contr. Univ. Michigan Herb. 14:153–203.

Roberts, A. 1983. A field guide to the sedges of the Cariboo Forest Region, British Columbia. Land Management Report No. 14. B.C. Ministry of Forests, Victoria, B.C., Canada.

Robertson, A.W. and F.C. Pollett. 1973. Checklist of the genus *Carex* in Newfoundland. Forest Research Centre, St. John's, Newfoundland.

Robertson, A.W. and F.C. Pollett. 1974. Illustrated key to the Carices in Newfoundland: based on floral structure. Forest Research Centre, St. John's, Newfoundland.

Savile, D.B.O. and J.A. Calder. 1953. Phylogeny of *Carex* in the light of parasitism by the smut fungi. Can. J. Bot. 31:164–174.

Schultze-Motel, W. 1968. Cyperaceen. *In* Hegi, Illustrierte Flora von Mitteleuropa II/1:81–160.

Schuyler, A.E. 1964. A biosystematic study of the *Scirpus cyperinus* complex. Proc. Acad. Nat. Sci. Philadelphia. 115:283–311.

Schuyler, A.E. 1967. A taxonomic revision of North American leafy species of *Scirpus*. Proc. Acad. Nat. Sci. Philadelphia 119:295–323.

Schuyler, A.E. 1971. Scanning electron microscopy of achene epidermis in species of *Scirpus* (Cyperaceae) and related genera. Proc. Acad. Nat. Sci. Philadelphia 123:29–52.

Schuyler, A.E. 1974. Typification and application of the names *Scirpus americanus* Pers., *S. olneyi* Gray, and *S. pungens* Vahl. Rhodora 76:51–52.

Scotter, G.W. and J.J. Hudson. 1975. *Carex illota* L.H. Bailey in Alberta. Can. Field-Naturalist 89:74–75.

Standley, L.A. 1985. Systematics of the Acutae group of *Carex* (Cyperaceae) in the Pacific Northwest. Syst. Bot. Mongr. 7:1–106.

Strandhede, S.O. 1960. A note on *Scirpus palustris* L. Bot. Not. 113:161–171.

Strandhede, S.O. 1966. Morphological variation and taxonomy in European *Eleocharis*, subser. Palustres. Opera Bot. 10(2):1–187.

Strandhede, S. 1967. Comments on some phytogeographical and biological problems in the complex of *Eleocharis palustris*. Bot. Not. 120:503–504.

Strandhede, S. 1967. *Eleocharis* subser. Eleocharis in North America. Taxonomical concepts and chromosome numbers. Bot. Not. 120:355–368.

Svenson, H.K. 1929, 1932, 1934, 1937, 1939. Monographic studies in the genus *Eleocharis*. Rhodora 31(1929):121–135, 152–163, 167–191, 199–219, 224–242; 34(1932):193–203, 215–227; 36(1934):377–389; 39(1937):210–231, 236–273; 41(1939):1–19, 43–77, 90–110.

Svenson, H.K. 1947. The group of *Eleocharis palustris* in North America. Rhodora 49:61–67.

Svenson, H.K. 1953. The *Eleocharis obtusa-ovata* complex. Rhodora 55:1–6.

Svenson, H.K. 1957. *Eleocharis*. North America Flora 18:509–540.

Svenson, H.K. 1957. Cyperaceae, Scirpeae. North America Flora 18:505–556.

Toivonen, H. 1974. Chromatographic comparison of the species *Carex* section Heleonastes and some *Carex canescens* hybrids in Eastern Fennoscandia. Ann. Bot. Fenn. 11:225–230.

Toivonen, H. 1981. Notes on the nomenclature and taxonomy of *Carex canescens* (Cyperaceae). Ann. Bot. Fenn. 18:91–97.

Toivonen, H. 1981. Spontaneous *Carex* hybrids of Heleonastes and related sections in Fennoscandia. Acta Bot. Fenn. 116:1–51.

Toivonen, H. and T. Timonen. 1976. Perigynium and achene epidermis in some species of *Carex* subgen. *Vignea* (Cyperaceae), studied by scanning electron microscopy. Ann. Bot. Fenn. 13:49–59.

Walter, S.K. 1975. A preliminary study of the achene epidermis of certain *Carex* (Cyperaceae) using scanning electron microscopy. Mich. Bot. 14:67–72.

Whitkus, R. 1981. A taxonomic investigation of the *Carex macloviana* D'Urv. aggregate in western Canada and Alaska. M. Sc. Thesis, Dept. Botany, Univ. of Alberta, Edmonton, Alta. 194 p.

IRIDACEAE

Bicknell, E.P. 1899. Studies in *Sisyrinchium*, IV. *S. angustifolium* and related species of the West and Northwest. Bull. Torrey Club 26:445:457.

Bicknell, E.P. 1900. Studies in *Sisyrinchium*, VII. The species of British America. Bull. Torrey Club 27:237–246.

Cholowa, A.F. and D.M. Henderson. 1991. Nomenclatural changes in *Sisyrinchium douglasii*. Madrono 38(4):232.

Cody, W.J. 1961. *Iris pseudoacorus* L. escaped from cultivation in Canada. Can. Field-Naturalist 75:139–142.

Cohn, V.A. 1967. A Guide to Pacific Coast Irises. British Iris Society. 39 p.

De Vries, B. 1966. *Iris missouriensis* Nutt. in southwestern Alberta and in northern central British Columbia. Can. Field-Naturalist 80:158–160.

Henderson, D.N. 1976. A biosystematic study of Pacific northwestern blue-eyed grasses (*Sisyrinchium*, Iridaceae). Brittonia 28:149–176.

Mosquin, T. 1970. Chromosome numbers and a proposal for classification in *Sisyrinchium* (Iridaceae). Madrono 20:269–275.

Raven, P.H. and J.H. Thomas. 1970. *Iris pseudoacorus* in western North America. Madrono 20:390–391.

JUNCACEAE

Catling, P.M. and K.W. Spicer. 1987. The perennial *Juncus* of section *Poiophylli* in the Canadian prairie provinces. Can. J. Bot. 65:750–760.

Coffey-Swab, J. 1993. Juncaceae, Rush Family. *In* Hickman, J.C. (ed.). The Jepson manual: higher plants of Califonica. Univ. CA Press, Berkeley and Los Angeles.

Ebinger, J.E. 1964. Taxonomy of the subgenus Pterodes of genus *Luzula*. Mem. New York Bot. Gard. 10:279–304.

Fernald, M.L. and K.M. Wiegand. 1910. The North American variation of *Juncus effusus*. Rhodora 12:81–93.

Fernald, M.L. 1924. *Juncus triglumis* and its American representative. Rhodora 26:201–203.

Hämet-Ahti, L. 1965. *Luzula piperia* (Cov.) M.E. Jones, an overlooked woodrush in western North America and eastern Asia. Aquilo. Ser. Botanica 3:11–21.

Hämet-Ahti, L. 1971. A synopsis of the species of *Luzula*, subgenus Anthelaea Griseb. (Juncaceae) indigenous in North America. Ann. Bot. Fenn. 8:368–381.

Hämet-Ahti, L. 1973. Notes on the *Luzula arcuata* and *L. parviflora* groups in eastern Asia and North America. Ann. Bot. Fenn. 12:27–29.

Hämet-Ahti, L. 1975. Additional notes on *Luzula subcongesta* and *L. parviflora* (Juncaceae) in North America. Ann. Bot. Fenn. 12:27–29.

Hämet-Ahti, L. 1980. *Juncus alpinoarticulatus*: the legitimate name for *Juncus alpinus*. Ann. Bot. Fenn. 17:341–342.

Hämet-Ahti, L. 1980. The *Juncus effusus* aggregate in eastern North America. Ann. Bot. Fenn. 17:183–191.

Hämet-Ahti, L. 1986. North America races of *Juncus alpinoarticulatus* (Juncaceae). Ann. Bot. Fenn. 23:277–281.

Hermann, F.J. 1964. The *Juncus mertensianus* complex in western North America. Leafl. West. Bot. 10:81–87.

Hermann, F.J. 1975. Manual of the rushes (*Juncus* spp.) of the Rocky Mountains and Colorado Basin. USDA Forest Service, General Technical Report RM-18. 107 p.

Hermann, F.J. and W.A. Weber. 1977. *Juncus triglumis* in North America. Rhodora 79:160–162.

Kirschner, J. 1990. *Luzula multiflora* and allied species (Juncaceae): a nomenclatural study. Taxon 39:106–114.

Nilsson, O. and S. Snogerup. 1971–1972. Drawings of Scandinavian plants, *Juncus* L. Bot. Not. 124:1–8, 179–186, 311–316, 436–441; 125:1–8, 131–138, 203–211.

Stuckey, R.L. 1980. The migration and establishment of *Juncus gerardii* (Juncaceae) in the interior of North America. Sida 8:334–347.

Van Loenhoud, P.J. and A.A. Sterk. 1976. A study of *Juncus bufonius* complex in the Netherlands. Acta Bot. Neerl. 25:193–204.

JUNCAGINACEAE

Löve, A. and D. Löve. 1958. Biosystematics of *Triglochin maritimum* Agg. Nat. Can. 85:156–165.

LEMNACEAE

Daubs, E.H. 1965. A monograph of Lemnaceae. Illinois Biol. Monogr. 34:1–118. Univ. Illinois Press, Urbana, Illinois.

de Lange, L. and A.H. Pieterse. 1973. A comparative study of the morphology of *Lemna gibba* L. and *Lemna minor* L. Acta. Bot. Neerl. 22:510–517.

den Hartog, C. and F. van der Plas. 1970. A synopsis of the Lemnaceae. Blumea 28:355–368

Dore, W.G. 1957. *Wolffia* in Canada. Can. Field-Naturalist 71:10–16.

Landolt, E. 1975. Morphological differentiation and geographical distribution of the *Lemna gibba-Lemna minor* group. Aqua. Bot. 1:345–363.

Landolt, E. 1980. Biosystematische untersuchungen in der familie der wasserlinsen (Lemnaceae). Veröff. Geobot. Inst. Eth. Siftung Rübel, Zürich: heft. 70.

McClure, J.S. and R.E. Alston. 1966. A chemotaxonomic study of Lemnaceae. Amer. J. Bot. 53:849–860.

LILIACEAE

Applegate, E.I. 1935. The genus *Erythronium*: a taxonomic and distributional study of the western North American species. Madrono 3:58–113.

Beetle, D.E. 1944. A monograph of the North American species of *Fritillaria*. Madrono 7:133–159.

Fassett, N.C. 1935. A study of *Streptopus*. Rhodora 37:88–113.

Fukuda, I. 1969. Chromosome compositions of natural populations in *Trillium ovatum*. Science Reports of Tokyo Women's Christ. Coll. 7–11:110–137.

Fukuda, I. and R.B. Channell. 1975. Distribution and evolutionary significance of chromosome variation in *Trillium ovatum.* Evolution 29:257–266.

Galway, D.H. 1945. The North American species of *Smilacina* (Liliaceae). Amer. Midl. Naturalist 33:644–666.

Gould, F.W. 1942. A systematic treatment of the genus *Camassia* Lindl. Amer. Midl. Naturalist 28:712–742.

Guppy, A.G. 1968. The rare *Trillium* of Vancouver Island. Amer. Rock Gard. Soc. Bull. 26:119–120.

Hitchcock, C.L. 1944. The *Tofieldia glutinosa* complex of western North America. Amer. Midl. Naturalist 31:487–498.

Jones, Q. 1951. A cytotaxonomic study of the genus *Disporum* in North America. Contr. Gray Herb. 173:1–39.

Kawano, S. and H.H. Iltis. 1966. Cytotaxonomy of the genus *Smilacina* (Liliaceae). II. Chromosome morphology and evolutionary consideration of New World species. Cytologia 31:12–28.

Kawano, S., M. Ihara and M. Suzuki. 1968. Biosystematic studies on *Maianthemum* (Liliaceae-Polygonatae). II. Geography and ecological life history. Jap. J. Bot. 20:35–66.

Kawano, S., M. Ihara, M. Suzuki and H.H. Iltis. 1967. Biosystematic studies on *Maianthemum* (Liliaceae-Polygonatae). I. Somatic chromosome number and morphology. Bot. Mag. Tokyo 80:345–352.

Kawano, S. and M. Suzuki. 1971. Biosystematic studies on *Maianthemum.* VI. Variation in gross morphology of *Maianthemum bifolium* and *M. canadense* with special reference to their taxonomic status. Bot. Mag. (Tokyo) 84:349–361.

Kawano, S., M. Suzuki and S. Kojima. 1971. Biosystematic studies in *Maianthemum* (Liliaceae-Polygonatae). V. Variation in gross morphology, karyology and ecology of North American populations of *M. dilatatum* sensu lato. Bot. Mag. (Tokyo) 84:299–318.

Kupchan, S.M., J.H. Zimmerman and A. Afonso. 1961. The alkaloids and taxonomy of *Veratrum* and other genera. Lloydia 24:1–26.

Marchant, C.J. and R.M. Macfarlane. 1980. Chromosome polymorphism in triploid population of *Fritillaria lanceolata* Pursh (Liliaceae) in California. Bot. J. Linn. Soc. (London) 81:135–154.

Maule, S.M. 1959. *Xerophyllum tenax*, squawgrass, its geographical distribution and its behaviour on Mount Rainier, Washington. Madrono 15:39–48.

McNeal, D.W. and M. Ownbey. 1973. Bulb morphology of some western North American species of *Allium.* Madrono 22:10–24.

Mitchell, R.J. 1970. The genus *Trillium* – II. J. Scott. Rock Gard. Club 12:16–20.

Niehaus, T.F. 1971. A biosystematic study of the genus *Brodiaea* (Amaryllidaceae). Univ. Calif. Publ. Bot. 60:1–66.

Ownbey, M. 1948. The identity and delimitation of *Allium tolmei* Baker. Madrono 9:233–238.

Ownbey, M. and H.C. Aase. 1955. Cytotaxonomic studies in *Allium.* I. The *Allium canadense* alliance. Res. Stud. State Coll. Wash. 23:1–106.

Reveal, J.L. 1971. A new combination in *Tofieldia glutinosa* (Liliaceae). Rhodora 73:53–55.

Taylor, T.M.C. 1966. The Lily family (Liliaceae) of British Columbia. B.C. Prov. Mus. Handbook 25, Victoria. 109 p.

Taylor, T.M.C. and A.F. Szczawinski. 1974. *Trillium ovatum* Pursh *forma hibbersonii* Taylor et Szczawinski *f. nov.* Syesis 7:250.

Turrill, W.B. and J.R. Sealy. 1980. Studies in the genus *Fritllaria* (Liliaceae). Hooker's Icones Plantarum 39 (I & II).

Wiley, L. 1973. Tetramerous *Trillium hibbersonii.* Amer. Rock Gard. Soc. Bull. 31:45–47.

ORCHIDACEAE

Ackerman, J.D. 1975. Reproductive biology of *Goodyera oblongifolia* (Orchidaceae). Madrono 23:191–198.

Ackerman, J.D. 1977. Biosystematics of the genus *Piperia* Rydb. (Orchidaceae). J. Linn. Soc. Bot. 75:245–270.

Correll, D.S. 1943. The genus *Habenaria* in western North America. Leafl. West. Bot. 3:233–256.

Correll, D.S. 1950. Native orchids of North America north of Mexico. Chronica Botanica Co., Waltham, Mass. 399 p.

Haley, P.L., J.D. Michaud and J. Arditti. 1980. Morphometry of orchid seeds. III. Native California and related species of *Goodyera, Piperia, Platanthera* and *Spiranthes*. Amer. J. Bot. 67:508–510.

Kallunki, J.A. 1981. Reproductive biology of mixed-species populations of *Goodyera* (Orchidaceae) in northern Michigan. Brittonia 33:137–155.

Kott, L. and A.E. Kott. 1974. A record of the orchid *Malaxis monophyllos* (L.) Sw. from northeastern British Columbia. Can. Field-Naturalist 88:89.

Luer, C.A. 1975. The native orchids of the United States and Canada excluding Florida. N.Y. Botanical Garden, New York. 361 p.

Morgan, R. and J.D. Ackerman. 1990. Two new Piperias (Orchidaceae) from Western North America. Lindleyana 5:205–211.

Revel, R.D. and K.I. Beamish. 1972. *Liparis loeselii* (L.) L.C. Rich., a new species for British Columbia. Syesis 5:287.

Schrenk, W.J. 1975. Zum Problem der Variabilität innerhalb der Gattung *Platanthera* L.C. Rich. (*Habenaria* Willd. sl.p.p.) in Nordamerika. Orchidee 26:258–263.

Szczawinski, A.F. 1959. The Orchids of British Columbia. B.C. Prov. Mus. Handbook 16, Victoria. 124 p.

Szczawinski, A.F. 1960. New records of Chamisso's orchid (*Habenaria chorisiana* Cham.) for British Columbia. Victoria Naturalist (Victoria) 17:35–36.

POACEAE

Alexeev, E.B. 1980. *Festuca* L. Subgenera et sectiones novae ex America Boreali et Mexica Nov. Sist. Vyssh. Rast. 17:42–53.

Alexeev, E.B. 1982. New and little-known fescues (*Festuca* L.) of North America. Byull. Mosk. Obva. Inspyt. Prir., Otd. Biol. 87:109–118.

Alexeev, E.B. 1985. *Festuca* L. (Poaceae) in Alaska et Canada. Nov. Syst. Pl. Vasc. 22:5–35.

Anderson, D.E. 1961. Taxonomy and distribution of the genus *Phalaris*. Iowa State Coll. J. Sci. 36:1–96.

Arnow, L.A. 1981. *Poa secunda* Presl versus *P. sandbergii* Vasey (Poaceae). Syst. Bot. 6:412–421.

Auquier, P. 1968. *Festuca rubra* L. subsp. *litoralis* (F.G.W. Mey.) Auquier: morphologie, ecologie, taxonomie. Bull. Jard. bot. nat. Belg. 38:181–192.

Auquier, P. 1971. *Festuca rubra* L. subsp. *pruinosa* (Hack.) Piper: morphologie, ecologie, taxonomie. Lejeunia, N.S. 56:1–16.

Auquier, P. 1971. *Festuca rubra* L. var. *tenuifolia* (How.) How. est synonyme de *F. rubra* L. subsp. *litoralis* (F.G.W. Mey.) Auquier. Bull. Jard. bot. nat. Belg. 41:289–292.

Auquier, P. 1971. Le probleme de *Festuca rubra* L. subsp. *arenaria* (Osb.) Richt. et de ses relations avec *F. juncifolia* St.-Amans. Lejeunia, N.S. 57:1–24.

Barkworth, M.E. 1978. A taxonomic study of the large-glumed species of *Stipa* (Gramineae) occurring in Canada. Can. J. Bot. 56:606–625.

Barkworth, M.E., J. McNeill and J. Maze. 1979. A taxonomic study of *Stipa nelsonii* Poaceae with a key distinguishing it from related taxa in western North America. Can. J. Bot. 57:2539–2554.

Baum, B.R. 1967. Kalm's specimens of North American grasses – their evaluation for typification. Can. J. Bot. 45:1845–1852.

Baum, B.R. 1968. Delimitation of the genus *Avena* (Gramineae). Can. J. Bot. 46:121–132.

Baum, B.R. 1968. On some relationships between *Avena sativa* and *A. fatua* (Gramineae) as studied from Canadian material. Can. J. Bot. 46:1013–1024.

Baum, B.R. 1969. The role of the lodicule and epiblast in determining natural hybrids of *Avena sativa X fatua* in cultivated oats. Can. J. Bot. 47:85–91.

Baum, B.R. 1974. Classification of the oat species (*Avena*, Poaceae) using various taximetric methods and an information-theoretic model. Can. J. Bot. 52:2241–2262.

Baum, B.R. 1977. Oats: wild and cultivated. A monograph of the genus *Avena* L. (Poaceae). Biosyst. Res. Inst. Monogr. No. 14, Ottawa. 463 p.

Baum, B.R. 1977. Taxonomy of the tribe Triticeae (Poaceae) using various numerical techniques. I. Historical perspectives, data accumulation, and character analysis. Can. J. Bot. 55:1712–1740.

Baum, B.R. 1978. Taxonomy of the tribe Triticeae (Poaceae) using various numerical techniques. III. Synoptic key to genera and synopses. Can. J. Bot. 56:374–385.

Baum, B.R. 1979. The genus *Elymus* in Canada – Bowden's generic concept and key reappraised and relectotypification of *E. canadensis*. Can. J. Bot. 57:946–951.

Baum, B.R. 1980. Multivariate morphometric relationships between *Hordeum jubatum* and *Hordeum brachyantherum* in Canada and Alaska. Can. J. Bot. 58:604–623.

Baum, B.R. and J.N. Findlay. 1973. Preliminary studies in the taxonomy of *Danthonia* in Canada. Can. J. Bot. 51:437–450.

Baum, B.R. and T. Rajhathy. 1976. A study of *Avena macrostachya*. Can. J. Bot. 54:2434–2439.

Beetle, A.A. 1943. The North American variations of *Distichlis spicata*. Bull. Torr. Bot. Club 70: 638–650.

Beetle, A.A. 1955. Wheatgrasses of Wyoming. Wyoming Agric. Exper. Stat. Bull. 336.

Böcher, T.W. 1959. Tetraploid and hexaploid *Trisetum spicatum*. A cytotaxonomic study. Bot. Tidskr. 55:23–29.

Boivin, B. 1948. Centurie de plantes canadiennes. Naturaliste Can. 75:79–84.

Boivin, B. and D. Love. 1960. *Poa agassizensis*, a new prairie bluegrass. Naturaliste Can. 87:173–180.

Bookman, P.A. 1980. Variation in *Bromus tectorum* (Poaceae) in eastern Washington. Madrono 27:36–42.

Bowden, W.M. 1957. Cytotaxonomy of sect. *Psammelymus* of the genus *Elymus* L. Can. J. Bot 35:951–993.

Bowden, W.M. 1958. Natural and artificial X *Elymordeum* hybrids. Can. J. Bot. 36:101–123.

Bowden, W.M. 1959. The taxonomy and nomenclature of the wheats, barleys, ryes and their wild relatives. Can. J. Bot. 37:657–684.

Bowden, W.M. 1959. Chromosome numbers and taxonomic notes on northern grasses. I. Tribe Triticeae. Can. J. Bot. 37:1143–1151.

Bowden, W.M. 1960. Chromosome numbers and taxonomic notes on northern grasses. II. Tribe Festuceae. Can. J. Bot. 38:117–131.

Bowden, W.M. 1960. Chromosome numbers and taxonomic notes on northern grasses. III. Twenty-five genera. Can. J. Bot. 38:541–557.

Bowden, W.M. 1960. The typification of *Elymus macounii* Vasey. Bull. Torrey Bot. Club 87:205–208.

Bowden, W.M. 1961. Chromosome numbers and taxonomic notes on northern grasses. IV. Tribe Festuceae: *Poa* and *Puccinellia*. Can. J. Bot. 39:123–137.

Bowden, W.M. 1962. Cytotaxonomy of the native and adventitive species of *Hordeum, Eremopyron, Secale, Sitanion*, and *Triticum* in Canada. Can. J. Bot. 40:1675–1711.

Bowden, W.M. 1964. Cytotaxonomy of the species and interspecific hybrids of the genus *Elymus* in Canada and neighboring areas. Can. J. Bot. 42:547–601.

Bowden, W.M. 1965. Cytotaxonomy of the species and interspecific hybrids of the genus *Agropyron* in Canada and neighboring areas. Can. J. Bot. 43:1421–1448.

Bowden, W.M. 1966. Citations of voucher specimens of the species and interspecific hybrids of the genus *Agropyron* in Canada and neighboring areas. Plant Res. Inst., Canada Dept. Agric., Ottawa. 31 p.

Bowden, W.M. 1967. Taxonomy of intergeneric hybrids of the tribe Triticeae from North America. Can. J. Bot. 45:711–724.

Bowden, W.M. and W.J. Cody. 1961. Recognition of *Elymus sibiricus* L. from Alaska and the District of Mackenzie. Bull. Torrey Bot. Club 88:153–155.

Boyle W.S. 1944. Cytological evidence for the taxonomic position of *Schizachne purpurascens*. Madrono 7:129–130.

Boyle, W.S. 1946. A cytotaxonomic study of the North American species of *Melica*. Madrono 8:1–26.

Boyle, W.S. 1963. A controlled hybrid between *Sitanion hystrix* and *Agropyron trachycaulum*. Madrono 17:10–16.

Boyle, W.S. and A.H. Holmgren. 1955. A cytotaxonomic study of natural and controlled hybrids between *Agropyron trachycaulum* and *Hordeum jubatum*. Genetics 40:539–545.

Boyle, W.S. and A.H. Holmgren. 1968. A cytotaxonomic study of a natural hybrid between *Agropyron cristatum* and *A. subsecundum*. Madrono 19:277–281.

Butters, F.K. and E.C. Abbe. 1947. The genus *Poa* in Cook County, Minnesota. Rhodora 49:1–21.

Callaghan, T.V. 1974. Intraspecific variation in *Phleum alpinum* L. with specific reference to polar populations. Arct. Alp. Res. 6:361–401.

Chapman, S.R. and L.J. Perry. 1973. Taxonomic and agronomic variation in *Agropyron spicatum* and *A. inerme*. J. Range Man. 26:41–43.

Chase, A. 1964. First Book of Grasses. The structure of grasses explained for beginners. 3rd edition. Smithsonian Institution Wash. D.C. 127 p.

Church, G.L 1949. A cytotaxonomic study of *Glyceria* and *Puccinellia*. Amer. J. Bot. 36:155–165.

Church, G.L. 1952. The genus *Torreyochloa*. Rhodora 54:197–200.

Church, G.L. 1967. Taxonomic and genetic relationships of eastern North American species of *Elymus* with setaceous glumes. Rhodora 69:121–162.

Clausen, R.T. 1952. Suggestion for the assignment of *Torreyochloa* to *Puccinellia*. Rhodora 54:42–45.

Clausen, J., W.M. Hiesey and M. Nobs. 1951. The *Poa* program. Carnegie Institute of Washington Year Book 53:152–156.

Clausen, J., P. Grun, A. Nygren and M. Nobs. 1951. Genetics and evolution of *Poa*. Carnegie Institute of Washington Year Book 53:109–111.

Clayton, W.D. 1968. The correct name of the common reed. Taxon 17:168–169.

Cody, W.J. 1967. *Elymus sibiricus* (Graminae) new to British Columbia. Can. Field-Natur. 81:275–276.

Cody, W.J., S.J. Darbyshire and C.F. Kennedy. 1990. A bluegrass, *Poa pseudoabbreviata* Roshev., new to the flora of Canada, and some additional records from Alaska. Can. Field-Natur. 104:589–591.

Collins, D.D. 1966. Natural hybrids of *Agropyron latiglume* and *A. scribneri*. Bull. Torrey Bot. Club 93:417–426.

Copple, R.F. and C.P. Pase. 1967. A vegetative key to some common Arizona range grasses. U.S. For. Serv. RM–27. U.S.D.A. 72 p.

Cotton, R. and C.A. Stace. 1976. Taxonomy of the genus *Vulpia* (Gramineae). I. Chromosome numbers and geographical distribution of the Old World species. Genetica 46:235–255.

Covas, G. 1949. Taxonomic observations on the North American species of *Hordeum*. Madrono 10:1–21.

Crampton, B. 1955. A new variety of *Stipa lemmonii*. Leafl. West. Bot. 7:220.

Crins, W.J. 1991. Paniceae. The genera of Paniceae (Gramineae: Panicoide) in the southeastern United States. J. Arnold. Arbor. Suppl. Ser. 1:177–312.

Darbyshire, S.J. 1984. *Sporobolus airoides (Eragrosteae:* Poaceae) in British Columbia. Syesis 17:11–12.

Darbyshire, S.J. and J. Cayoette. 1989. The biology of Canadian weeds. 92. *Danthonia spicata* (L.) Beauv. in Roem & Schult. Can. J. Plant Sci. 69:1217–1233.

Daubenmire, R. 1960. An experimental study of variation in the *Agropyron spicatum-A. inerme* complex. Bot. Gaz. 122:104–108.

Davis, J.I. and P.S. Manos. 1991. Isozyme variation and species delimitation in the *Puccinellia nuttalliana* complex (Poaceae): an application of the phylogenetic species concept. Syst. Bot. 16:431–445.

Dedecca, D.M. 1954. Studies on the California species of *Stipa* (Gramineae). Madrono 12:129–139.

DeLisle, D.G. 1963. Taxonomy and distribution of the genus *Cenchrus*. Iowa State Coll. J. Sci. 37:259–351.

Dewey, D.R. 1963. Natural hybrids of *Agropyron trachycaulum* and *A. scribneri*. Bull. Torrey Bot. Club 90:111–122.

Dewey, D.R. 1964. Natural and synthetic hybrids of *Agropyron spicatum X Sitanion hystrix*. Bull. Torrey Bot. Club 91:396–405.

Dewey, D.R. 1974. Cytogenetics of *Elymus sibiricus* and its hybrids with *Agropyron tauri*, *Elymus canadensis*, and *Agropyron caninum*. Bot. Gaz. 135:80–87.

Dewey, D.R. 1976. Cytogenetics of *Agropyron pringlei* and its hybrids with *A. spicatum, A. scribneri, A. violaceum* and *A. dasystachyum*. Bot. Gaz. 137:179–185.

Dewey, D.R. and A.H. Holmgren. 1962. Natural hybrids of *Elymus cinereus X Sitanion hystrix*. Bull. Torrey Bot. Club 89:217–228.

Dobrenz, A.K. and A.A. Beetle. 1966. Cleistogenes in *Danthonia*. J. Range Man. 19:292–296.

Dore, W.G. 1950. Supposed natural hybrid between *Agropyron* and *Hystrix*. Can. Field-Natur. 64:39–40.

Dore, W.G. and J. McNeill. 1980. Grasses of Ontario. Can. Dept. Agric. Res. Branch Monogr. 26. 566 p.

Erdman, K.S. 1965. Taxonomy of the genus *Sphenopholis* (Gramineae). Iowa State Coll. J. Sci. 39:289–336.

Evans, G.R. and E.W. Tisdale. 1972. Ecological characteristics of *Aristida longiseta* and *Agropyron spicatum* in west-central Idaho. Ecology 53:137–142.

Fassett, N.C. 1946. *Glyceria pallida* and *G. fernaldii*. Bull. Torrey Bot. Club 73:463–464.

Fernald, M.L. 1934. *Agropyron trachycaulum* versus *A. pauciflorum*. Rhodora 36:417–420.

Fernald, M.L. 1943. Contr. Gray Herb. Harvard Univ. No. CXLVIII. II. Notes on Danthonia. Rhodora 45:239–246.

Fernald, M.L. & C.A. Weatherby. 1916. The genus *Puccinellia* in eastern North America. Rhodora 18:1–23.

Findlay, J.N. and B.R. Baum. 1974. The nomenclatural implications of the taxonomy of *Danthonia* in Canada. Can. J. Bot. 52:1573–1582.

Frederiksen, S. 1977. The *Festuca brachyphylla* group in Greenland. Bot. Not. 130:269–277.

Frederiksen, S. 1979. *Festuca minutiflora* Rydb., a neglected species. Bot. Not. 132:315–318.

Frederiksen, S. 1983. *Festuca brachyphylla, F. saximontana* and related species in North America. Nord. J. Bot. 2–1982:525–536.

Gillett, J.M. and H.A. Senn. 1960. Cytotaxonomy and intraspecific variation of *Agropyron smithii* Rydb. Can. J. Bot. 38:747–760.

Gould, F.W. 1947. Nomenclatural changes in *Elymus* with a key to the Californian species. Madrono 9:120–128.

Gould, F.W. 1967. The grass genus *Andropogon* in the United States. Brittonia 19:70–76.

Gould, F.W. 1968. Grass systematics. McGraw-Hill Book Co., New York. 382 p.

Gould, F.W., M.A. Ali and D.E. Fairbrothers. 1972. A revision of *Echinochloa* in the United States. Amer. Midl. Naturalist 87:36–59.

Gould, F.W. and C.A. Clark. 1978. *Dichanthelium* (Poaceae) in the United States and Canada. Ann. Missouri Bot. Gard. 65(4):1088–1132.

Greene, C.W. 1980. The systematics of *Calamagrostis* (Gramineae) in eastern North America. Ph.D. Thesis, Harvard Univ. Cambridge, Mass.

Guedes, M. and P. Dupuy. 1976. Comparative morphology of lodicules in grasses. J. Linn. Soc. Bot. 73:317–331.

Hackel, E. 1882. Monographia *Festuca* Europaearum. Kassel und Berlin, Th.Fisher. 216 p. [Reprint 1964, Micromethods Ltd.].

Hagerup, O. 1939. Studies on the significance of polyploidy. III. *Deschampsia* and *Aira*. Hereditas 25:185–192.

Hall, B.M. 1955. Genetic analysis of interspecific hybrids in the genus *Bromus*, sect. *Ceratochloa*. Genetics 40:175–192.

Harbera, D.H. 1962. Some observations on natural clones in *Festuca ovina*. New Phytol. 61:85–100.

Harlan, J.R. 1945. Cleistogamy and chasmogamy in *Bromus carinatus* Hook & Arn. Amer. J. Bot. 32:66–72.

Henrard, J.T. 1937. A study in the genus *Vulpia*. Blumea 2:299–326.

Henrad, J.T. 1950. Monograph of the genus *Digitaria*. Universitare Pers, Leiden.

Herzman, C.W., A.G. Everson, M.H. Mickey, I.R. Porter, R.H. Searway and C.S. Fonte. 1959. Handbook of Colorado Native Grasses. Bull. 450-A Colorado State Univ. Extension Service, Fort Collins.

Hitchcock, C.L. 1973. Poaceae. *In* C.L. Hitchcock and A. Cronquist (eds.) Flora of the Pacific Northwest. Univ. Wash. Press, Seattle. 730 p.

Hitchcock. A.S. 1925. The North American species of *Stipa*. Contr. U.S. Natl. Herb. 24:215–262.

Hitchcock, A.S. 1971. Manual of Grasses of the United States. Vol. 1 and Vol. 2, 2nd. edition. General Publishing Co. Ltd., Toronto. (Dover Public.) 1051 p.

Hitchcock, A.S. and A. Chase. 1910. The North American species of *Panicum*. Contr. U.S. Natl. Herb. 15:1–396.

Hitchcock, C.L. and R.W. Spellenberg. 1968. A new *Oryzopsis* from Idaho. Brittonia, 20:162–165.

Hodgkinson, H.S. and A.E. Young. 1973. Rough fescue (*Festuca scabrella*) in Washington. J. Range Man. 26:25–26.

Holmen, K. 1964. Cytotaxonomical studies in the arctic Alaskan flora. The genus *Festuca* (Gramineae). Bot. Not. 117:109–118.

Hubbard, C.E. 1942. The collective species *Glyceria fluitans*. J. Ecol. 30:233.

Hubbard, W.A. 1969. The Grasses of British Columbia. B.C. Prov. Mus. Handbook No. 9, Victoria. 205 p.

Hultén, E. 1959. The *Trisetum spicatum* complex. *Trisetum spicatum* (L.) Richt., an arctic-montane species with world-wide range. Svensk Bot. Tidskr. 53:203–228.

Hunziker, J.H. 1955. Artificial and natural hybrids in the Gramineae, tribe Hordeae. VIII. Four hybrids of *Elymus* and *Agropyron*. Amer. J. Bot. 42:459–467.

Jamison, D.W. 1979. *Poa kelloggii* Vasey, new for British Columbia. Syesis 12:177.

Johnson, B.L. 1945. Natural hybrids between *Oryzopsis hymenoides* and several species of *Stipa*. Amer. J. Bot. 32:599–608.

Johnson, B.L. 1962. Amphiploidy and introgression in *Stipa*. Amer. J. Bot. 49:253–262.

Johnston, A. and H.E. Crosby. 1966. Rhizomatous forms of *Festuca scabrella*. Can. J. Bot. 46:211–212.

Johnston, A. and M.D. MacDonald. 1967. Floral initiation and seed production in *Festuca scabrella* Torr. Can. J. Plant Sci. 47:577–583.

Jones, E.K. and N.C. Fassett. Subspecific variation in *Sporobolus cryptandrus*. Rhodora 52:125–126.

Jones, K. 1960. The typification of the genus *Agropyron* Gaertn. Taxon 9:55–56.

Kam, Y.K. and J. Maze. 1974. Studies on the relationships and evolution of supraspecific taxa utilizing developmental data: 2. Relationships and evolution of *Oryzopsis hymenoides*, *O. virescens*, *O. kingii*, *O. micrantha* and *O. asperifolia*. Bot. Gaz. 135:227–247.

Kawano, S. 1963. Cytogeography and evolution of the *Deschampsia caespitosa* complex. Can. J. Bot. 41:719–742.

Kawano, S. 1965. *Calamagrostis purpurascens* R. Br. and its identity. Acta Phytotax. Geobot. 21:73–89.

Kellogg, E.A. 1983. A biosystematic study of the *Poa secunda* complex. Ph.D. Thesis, Harvard Univ., Cambridge, Mass.

Kellogg, E.A. 1985. A biosystematic study of the *Poa secunda* complex. J. Arnold Arbor. 66:201–242.

Kerguelen, M. 1975. Les Gramineae (Poaceae) de la flore francaise essai de mise au point taxonomique et nomenclaturale. Lejeunia N.S. No. 75:1–343.

Kjellqvist, E. 1964. *Festuca arenaria* Osb., a misinterpreted species. Bot. Not. 117:389–396.

Klemmedson, J.O. and J.G. Smith. 1964. Cheatgrass (*Bromus tectorum* L.). Bot. Rev. 30:226–262.

Koch, S.D. 1974. The *Eragrostis pectinacea – pilosa* complex in North and Central America (Gramineae: Eragrostoideae). Illinois Biol. Monogr. 48:1–74.

Koyama, T. and S. Kawano. 1964. Critical taxa of grasses with North American and eastern Asiatic distribution. Can. J. Bot. 42:859–884.

Lawrence, W.E. 1945. Some ecotypic relations of *Deschampsia caespitosa*. Amer. J. Bot. 32:298–314.

Lonard, R.I. 1970. A biosystematic study of the genus *Vulpia* (Gramineae).Ph.D. Thesis, Texas A&M Univ., College Station, Texas.

Lonard, R.I. and F.W. Gould. 1974. The North American species of *Vulpia* (Gramineae). Madrono 22:217–230.

Marchand, L.S. and A. MacLean. 1965. Observations on vegetative proliferation in *Agropyron spicatum*. Can. J. Bot. 43:1533–1537.

Marsh, B.L. 1952. A taxonomic revision of the genus *Poa* in the United States and southern Canada. Amer. Midl. Natur. 47:202–250.

Mitchell, W.W. 1967. On the *Hordeum jubatum-H. brachyantherum* question. Madrono 19:108–110.

Mitchell, W.W. 1967. Taxonomic synopsis of *Bromus* section Bromopsis (Gramineae) in Alaska. Can. J. Bot. 45:1309–1313.

Mitchell, W.W. 1968. Taxonomy, variation, and chorology of three chromosome races of the *Calamagrostis canadensis* complex in Alaska. Madrono 19:235–246.

Mitchell, W.W. 1968. Hybridization within the Triticeae of Alaska: a new x *Elyhordeum* and comments. Rhodora 70:467–473.

Mitchell, W.W. and R.W. Pohl. 1966. Variation and aneuploidy in *Muhlenbergia glomerata*. Amer. Midl. Natur. 76:211–221.

Mitchell, W.W. and A.C. Wilton. 1964. The *Hordeum jubatum-caespitosum- brachyantherum* complex in Alaska. Madrono 17:269:280.

Mitchell, W.W. and A.C. Wilton. 1965. Redefinition of *Bromus ciliatus* and *B. richardsonii* in Alaska. Brittonia 17:278–288.

Mitchell, W.W. and A.C. Wilton. 1966. A new tetraploid Brome, section Bromopsis, of Alaska. Brittonia 18:162–166.

Mobberly, D.G. 1956. Taxonomy and distribution of the genus *Spartina*. Iowa State Coll. J. Sci. 30:471–574.

Morrison, M.E.S. 1959. Interrelationships of some of the species of *Trisetum* from North America and Europe. Can. J. Genet. Cytol. 1:84–88.

Nannfeldt, J.A. 1940. On the polymorphy of *Poa arctica*. R. Br. Symp. Bot. Upsaliensis 4:1.

Nath, J. 1967. Cytogenetical and related studies in the genus *Phleum* L. Euphytica 16:267–282.

Nygren, A. 1946. The genesis of some Scandinavian species of *Calamagrostis*. Hereditas 32:131–262.

Nygren, A. 1949. Studies on vivipary in the genus *Deschampsia*. Hereditas 35:27–32.

Nygren, A. 1954. Investigations on North American *Calamagrostis*. I. Hereditas 40:377–397.

Nygren, A. 1958. Investigations on North American *Calamagrostis*. II. Lantbrukshögskolans Ann. 24:363–368.

Passey, H.B. and V.K. Hugie. 1963. Variation in bluebunch wheatgrass in relation to environment and geographic location. Ecology 44:158–161.

Pavlick, L.E. 1983a. The taxonomy and distribution of *Festuca idahoensis* in British Columbia and northwestern Washington. Can. J. Bot. 61:345–353.

Pavlick, L.E. 1983b. Notes on the taxonomy and nomenclature of *Festuca occidentalis* and *F. idahoensis*. Can. J. Bot. 61:337–344.

Pavlick, L.E. 1983c. *Festuca viridula* (Poaceae): Reestablishment of its original lectotype. Taxon 32:117–120.

Pavlick, L.E. 1984. Studies on the *Festuca ovina* complex in the Canadian Cordillera. Can. J. Bot. 62:2448–2462.

Pavlick, L.E. 1985. A new taxonomic survey of the *Festuca rubra* complex in northwestern North America, with emphasis on British Columbia. Phytologia 57:1–17.

Pavlick, L.E. and J. Looman. 1984. Taxonomy and nomenclature of rough fescues, *Festuca altaica, F. campestris* (*F. scabrella* var. *major*), and *F. hallii*, in Canada and the adjacent part of the United States. Can. J. Bot. 62:1739–1749.

Philipson, W.R. 1937. A revision of British species of the genus *Agrostis*. J. Linn. Soc. (London) Bot. 51:73–151.

Pilger, R. 1954. Das system der Gramineae. Bot. Jahrb. Syst. 76:281–384.

Piper, C.V. 1906. North American species of *Festuca*. Contr. U.S. Natl. Herb. 10:1–48.

Pohl, R.W. 1969. *Muhlenbergia* subgenus *Muhlenbergia* (Gramineae) in North America. Amer. Midl. Naturalist 82:512–542.

Pohl, R.W. and W.W. Mitchell. 1964. Cytogeography of the rhizomatous American species of *Muhlenbergia*. Brittonia 17:107–112.

Pyrah, G.L. 1969. Taxonomic and distributional studies on *Leersia* (Gramineae). Iowa State Coll. J. Sci. 44:215–270.

Rajhathy, T. 1961. Cytogenetic studies in the genus *Hordeum*, V. *Hordeum jubatum* and the New World species. Can. J. Gen. Cytol. 3:378–390.

Rajhathy, T. 1966. Notes on the *Hordeum jubatum* complex. Madrono 18:243–244.

Rajhathy, T. and J.W. Morrison. 1959. Cytogenetic studies in the genus *Hordeum*. IV. Hybrids of *Hordeum jubatum, H. brachyantherum, H. vulgare* and a hexaploid *Hordeum* sp. Can. J. Gen. Cytol. 1:124–132.

Rebristaya, O.V. (ed.). 1964. Semeistvo Gramineae. *In* Tolmachev, A.I. (ed.) Arkticheskaya flora SSSR. vol. II. Nauka, Moskva. 272 p.

Reeder, J.R. 1953. Affinities of the grass genus *Beckmannia* Host. Bull. Torrey Bot. Club. 80:187–196.

Reeder, J.R. and M.A. Ellington. 1960. *Calamovilfa*, a misplaced genus of the Gramineae. Brittonia 12:71–77.

Roberts, A.L. 1981. Common grasses of the Cariboo Forest Region. B.C. Ministry of Forests, Victoria. 29 p.

Rominger, J.M. 1962. Taxonomy of *Setaria* (Gramineae) in North America. Illinois Biol. Monogr. 29:1–132.

Runemark, H. and W.K. Heneen. 1968. *Elymus* and *Agropyron*, a problem of genera delimitation. Bot. Not. 121:51–80.

Saint-Yves, A. 1925. Contributions a l'etude des *Festuca* (Subgen. Eu-*Festuca*) de l'Amerique du Nord et du Mexique. Candollea 2:229–316.

Schmoll, H.M. 1939. A realignment of the *Panicum thermale* group. Madrono 5:90–96.

Scholz, H. 1971. Zur Systematik der Gattung *Bromus* L. subgenus *Bromus* (Gramineae). Willdenowia 6:139–159.

Schulz-Schaeffer, J. and P. Jurasits. 1962. Biosystematic investigations in the genus *Agropyron*. I. Cytological studies of species karyotypes. Amer. J. Bot. 49:940–953.

Scribner, F.L. and E.D. Merrill. 1910. The grasses of Alaska. U.S. Natl. Herb. Contr. 13:47–92.

Shinners, L.H. 1956. Illegitimacy of Persoon's species of *Koeleria* (Gramineae). Rhodora 58:93–96.

Smith, P. 1968. The *Bromus mollis* aggregate in Britain. Watsonia 6:327–344.

Snyder, L.A. 1950. Morphological variability and hybrid development in *Elymus glaucus*. Amer. J. Bot. 37:628–636.

Snyder, L.A. 1951. Cytology of inter-strain hybrids and the probable origin of variability in *Elymus glaucus*. Amer. J. Bot. 38:195–202.

Soreng, R.J. 1985. *Poa* L. in New Mexico, with a key to middle and southern Rocky Mountain species (Poaceae). Great Basin Natur. 45:395–422.

Soreng, R.J. 1986 Distribution and evolutionary significance of apomixis in diclinous *Poa* of western North America. Ph.D. Thesis, New Mexico State Univ., Las Cruces.

Soreng, R.J. 1991. Notes on new infraspecic taxa and hybrids of North American *Poa* (Poaceae). Phytologia 71:390–413.

Soreng, R.J. 1991. Systematics of the "Epiles" group of *Poa* (Poaceae). Syst. Bot. 16:507–528.

Sorensen, T. 1953. A revision of the Greenland species of *Puccinellia* Parl. with contributions to our knowledge of the arctic *Puccinellia* flora in general. Medd. Gronl. 136:1–179.

Spellenberg, R.W. 1969. Notes on *Oryzopsis hendersonii* (Gramineae). Madrono 19:283–286.

Spellenberg, R.W. 1970. *Panicum shastense* (Gramineae), a sterile hybrid between *P. pacificum* and *P. scribnerianum*. Brittonia 22:154–162.

Spellenberg, R.W. 1975. Synthetic hybridization and taxonomy of western North American *Dichanthelium*, group *Lanuginosa* (Poaceae). Madrono 23:134–153.

Spellenberg, R.W. 1975. Autogamy and hybridization as evolutionary mechanisms in *Panicum*, subgenus *Dichanthelium* (Gramineae). Brittonia 27:87–95.

St. John, H. 1941. The status of *Poa* secunda and of *Poa sandbergii* (Gramineae) in North America. New and noteworthy northwestern plants, Part 8. Amer. J. Bot. 28:78–81.

Stebbins, G.L., Jr. 1930. A revision of some North American species of *Calamagrostis*. Rhodora 32:35–57.

Stebbins, G.L., Jr. 1947. The origin of the complex of *Bromus carinatus* and its phytogeographic implications. Contr. Gray Herb. 165:42–55.

Stebbins, G.L., Jr. 1956. Cytogenetics and the evolution of the grass family. Amer. J. Bot. 43:890–905.

Stebbins, G.L., Jr. and B. Crampton. 1961. A suggested revision of the grass genera of temperate North America. Recent Advances Bot. 1:133–145.

Stebbins, G.L., Jr. and H.A. Tobgy. 1944. The cytogenetics of hybrids in *Bromus*. I. Hybrids within the section Ceratochloa. Amer. J. Bot. 31:1–11.

Stebbins, G.L., Jr. and A. Vaarama. 1954. Artificial and natural hybrids in the Gramineae, tribe Hordeae. VII. Hybrids and allopolyploids between *Elymus glaucus* and *Sitanion* spp. Genetics 39:378–395.

Stebbins, G.L., Jr., J.I. Valencia and R.M. Valencia. 1946. Artificial and natural hybrids in the Gramineae, tribe Hordeae. *Elymus*, *Sitanion*, and *Agropyron*. Amer. J. Bot. 33:338–351.

Stebbins, G.L., Jr. and D. Zohary. 1959. Cytogenetic and evolutionary studies in the genus *Dactylis*. Univ. Calif. Publ. Bot. 31:1–40.

Swallen, J.R. 1944. The Alaskan species of *Puccinellia*. J. Wash. Acad. Sci. 34:16–23.

Tateoka, T. 1957. Miscellaneous papers on the phylogeny of Poaceae (10). Proposition of a new phylogenetic system of Poaceae. J. Jap. Bot. 32:275–287.

Tateoka, T. 1974. A cytotaxonomic study of the *Calamagrostis purpurea- langsdorfii-canadensis* complex in the lowlands of Hokkaido. Bot. Mag. (Tokyo) 87:237–251.

Teeri, J.A. and L.G. Stowe. 1976. Climatic patterns and the distribution of C_4 grasses in North America. Oecologia 23:1–12.

Terrel, E.E. 1968. A taxonomic revision of the genus *Lolium*. Techn. Bull. USDA 1392:1–65.

Thieret, J.W. 1966. Synopsis of genus *Calamovilfa* (Gramineae). Castaneae 31:145–152.

Tzvelev, N.N. 1976. Zlaki SSSR. Nauka, Leningrad. 788 p.

Wagnon, H.K. 1952. A revision of the genus *Bromus*, sect. Bromopsis, of North America. Brittonia 7:415–480.

Weimarck, G. 1971. Variation and taxonomy of *Hierochloe* (Gramineae) in the northern hemisphere. Bot. Not. 124:129–175.

Widen, K.-G. 1971. The genus *Agrostis* L. in eastern Fennoscandia. Taxonomy and distribution. Flora Fennica 5:1–209.

Wilson, F.D. 1963. Revision of *Sitanion* (Triticeae, Gramineae). Brittonia 15:303–323.

POTAMOGETONACEAE

Aalto, M. 1970. Potamogetonaceae fruits: I. Recent and subfossil endocarps of the Fennoscandian species. Acta Bot. Fenn. 88:1–85.

Aalto, M. 1974. Potamogetonaceae fruits. II. *Potamogeton robbinsii*, a seldom fruiting North American pondweed species. Ann. Bot. Fenn. 11:29–33.

Catling, P.M. & I. Dobson. 1985. The biology of Canadian weeds. 69. *Potamogeton crispus* L. Can. J. Plant. Sci. 65:655–668.

Dandy, J.E. and G. Taylor. 1938. Studies of British Potamogetons. I. The typification of *Potamogeton pusillus*. J. Bot. 76:89–92.

Fernald, M.L. 1932. The linear-leaved North American species of *Potamogeton*, section *Axillares*. Mem. Amer. Acad. Arts, Ser. 2. 17(1):1–183. (Mem. Gray Herb. 3).

Haynes, R.R. 1974. A revision of North American *Potamogeton* subsection *Pusilli* (Potamogetonaceae). Rhodora 76:564–649.

Haynes, R.R. 1978. The Potamogetonaceae in the southeastern United States. (Generic flora of the southeastern United States.) J. Arnold Arbor. 59:170–191.

Haynes, R.R. 1986. Typification of Linnaean species of *Potamogeton* (Potamogetonaceae). Taxon 35:563–573.

Haynes, R.R. and J.L. Reveal. 1973. A re-evaluation of *Potamogeton fibrillosus* Fern. (Potamogetonaceae). Rhodora 75:75–77.

Hellquist, C.B. 1980. Correlation of alkalinity and the distribution of *Potamogeton* in New England. Rhodora 82:331–344.

Klekowski, E.J. and E.O. Beal. 1965. A study in the variation in the *Potamogeton capillaceus-diversifolius* complex (Potamogetonaceae). Brittonia 17:175–181.

Les, D.H. 1983. Taxonomic implications of aneuploidy and polyploidy in *Potamogeton* (Potamogentonaceae). Rhodora 85:301–324.

Ogden, E.C. 1943. The broad-leaved species of *Potamogeton* of North America north of Mexico. Rhodora 45:57–105, 119–163, 171–214.

Ogden, E.C. 1974. *Potamogeton* in New York. Bull. No. 423, New York State Museum and Science Service, Albany, N.Y. 20 p.

Reznicek, A.A. and R.S.W. Bobbette. 1976. The taxonomy of *Potamogeton* subsection *Hybridi* in North America. Rhodora 78:650–673.

Riley, J.L. 1979. Ontario's pondweeds (genus *Potamogeton*). Ont. Field Biol. 33:1–25.

Sorsa, P. 1988. Pollen morphology of *Potamogeton* and *Groenlandia* (Potamogetonaceae) and its taxonomic significance. Ann. Bot. Fenn. 25:179–199.

St. John, H. 1916. A revision of the North American species of *Potamogeton* of the section *Coleophylli*. Rhodora 18:121–138.

Stuckey, R.L. 1979. Distributional history of *Potamogeton crispus* (Curly pondweed) in North America. Bartonia 46:22–42.

Weigleb, B. 1990. The importance of stem anatomical characters for the systematics of the genus *Potamogeton* L. Flora 184:197–208.

RUPPIACEAE

Dandy, J.E. 1980. In Tutin, T.G. et. al. (eds.). Flora Europaea. Vol. 5. Cambridge Univ. Press, London.

Fernald, M.L. and K.M. Wiegand. 1914. The genus *Ruppia* in eastern North America. Rhodora 16:119–127.

Graves, A.H. 1908. The morphology of *Ruppia maritima*. Trans. Connecticut Acad. Arts 14:59–170.

Morong, T. 1893. The Naiadaceae of North America. Mem. Torrey Bot. Club 3:1–65.

Posluszny, U. and R. Sattler. 1974. Floral development of *Ruppia maritima* var. *maritima*. Can. J. Bot. 52:1607–1612.

Reise, G. 1962. Zur intragenerischen Taxonomie der Gattung *Ruppia* L. Z. Bot. 50:237–264.

Setchell, W.A. 1946. The genus *Ruppia* L. Proc. Calif. Acad. Sci. Ser. 4, 25:469–478.

Taylor, N. 1909. *Ruppia*. North American Flora 17:13–14.

SPARGANIACEAE

Cook, C.D.K. and M.S. Nicholls. 1986. A monographic study of the genus *Sparganium* (Sparganiaceae). Part 1. Subgenus *Xanthosparganium* Holmberg. Bot. Helvet. 96:213–267.

Engelmann, G. 1867. *Sparganium. In* Manual of the botany of the northern United States. 5th ed. Ivison, Blakeman, Taylor & Co. 703 p.

Harms, V.L. 1973. Taxonomic studies of North American *Sparganium*. I. *S. hyperboreum* and *S. minimum*. Can. J. Bot. 51:1629–1641.

Morong, T. 1988. Studies in the Typhaceae. II. *Sparganium*. Bull. Torrey Club 15:73–81.

Packer, J.G. 1983b. Flora of Alberta: *Sparganium angustifolium* and *Erigeron trifidus*. Can. J. Bot. 61:359–366.

Reveal, J.L. 1970. *Sparganium simplex* Huds., A superfluous name. Taxon 19:796–797.

ZOSTERACEAE

Bigley, R.E. and J.L. Barreca. 1982. Evidence for synonymizing *Zostera americana* den Hartog with *Zostera japonica* Aschers. and Graebn. Aqua. Bot. 14:349–356.

den Hartog, C. 1970. The sea-grasses of the world. North Holland, Publ. Co., Amsterdam.

Harrison, P.G. 1976. *Zostera japonica* (Aschers. and Graebn.) in British Columbia, Canada. Syesis 9:359–360

Phillips, R.C. 1970. Ecological notes on *Phyllospadix* (Potamogetonaceae) in the Northeast Pacific. Aqua. Bot. 6:159–170.

Phillips, R.C. and R.F. Shaw. 1976. *Zostera noltii* Hornem. in Washington, U.S.A. Syesis 9:355–358.

Setchell, W.A. 1929. Morphological and phenological notes on *Zostera marina* L. Univ. Calif. Publ. Bot. 14:420–421.

Setchell, W.A. 1933. A preliminary survey of the species of *Zostera*. Proc. Nat. Acad. U.S. 19:810–817.

APPENDIX

EXCLUDED SPECIES

HYDROCHARITACEAE

***Vallisneria spiralis* L.** This species, cited by Taylor and MacBryde, apparently is absent from BC (Brayshaw 1985) and Canada (Scoggan 1978). Our material is referrable to *V. americana*.

IRIDACEAE

Iris setosa* Pall. ex Link in Spreng. *et al. Cited in Taylor & MacBryde (1977) and Straley *et al.* (1985), but all specimens referred to as this species are *Iris missouriensis*. However, it may yet be found in NW BC.

JUNCACEAE

***Juncus canadensis* Gay.** This species, reported by Ceska (1975), was based on a mis-identification.

***Juncus marginatus* Rostk.** Reported for BC by Coffey-Swab (1993), but no specimens seen.

***Juncus orthophyllus* Coville** Cited by Scoggan (1978) this species was based on mis-identified specimens.

LEMNACEAE

***Lemna gibba* L.** Reported by Taylor and MacBryde (1977) but no material seen.

LILIACEAE

***Narcissus jonquilla* L.** Cited in Taylor and MacBryde (1977), but no specimens seen. A garden ornamental, not known to persist.

***Nothoscordum bivalve* (L.) Britt.** Cited in Taylor & MacBryde (1977) based on 1921 record from Victoria. No recent collections known.

***Scilla nonscripta* (L.) Hoffmanns. & Link.** Reported by Taylor & MacBryde (1977) but probably based on misidentifications of *Hyacinthoides hispanica (Scilla hispanica)*.

Smilacina racemosa* (L.) Desf. var. *racemosa The eastern North American variety cited in Taylor & MacBryde (1977) is not expected in BC.

ORCHIDACEAE

***Cypripedium fasciculatum* Kell. ex S. Wats.** Cited in Taylor & MacBryde (1977) but no specimens seen.

Malaxis monophyllos* var. *monophyllos Cited in Straley *et al.* (1985) but referrable to var. *brachypoda*. The nominate variety is European.

POACEAE

Agrostis canina* L. ssp. *canina Cited by Taylor and MacBryde (1977) and Scoggan (1978), this taxon is used in lawns and on golf courses and escapes occasionally but does not persist.

***Agrostis melaleuca* (Trin.) A.S. Hitchc.** (*A. alascana* Hult.?) These taxa are of uncertain application. They have generally been applied (i.e., in Taylor and MacBryde 1977, Scoggan 1978) to *A. mertensii* and northern forms of the taxon treated in this manual as *A. idahoensis*.

***Agrostis thurberiana* A.S. Hitchc.** (*Podagrostis thurberiana* [A.S. Hitchc.]) This species is reported from BC in most floras (e.g., Hitchcock *et al.* 1969, Taylor and MacBryde 1977, Scoggan 1978) but all specimens examined during this study were either *A. aequivalvis* or more vigorous specimens of *A. humilis*.

***Alopecurus myosuroides* Huds.** (*A. agrestis* L.) First reported by Hubbard (1969) and taken up by MacBryde and Taylor (1977), this species was never established in our area.

***Alopecurus saccatus* Vasey** (*A. howellii* Vasey) First reported by Henry (1915) and taken up by Taylor and MacBryde (1977) but probably based on collections of *A. carolinianus* (Scoggan 1978).

***Briza media* L.** Reported by Hubbard (1969) and Scoggan (1978) but not collected since 1917.

***Briza minor* L.** Cited by Jepson (1951) and Boivin (1966–1967) and taken up by Taylor and MacBryde (1977), this species does not persist in our area.

***Bromus arvensis* L.** Cited by Scoggan (1978) but no specimens seen.

***Bromus erectus* Huds.** (*B. macounii* Vasey). Reported from the Victoria area by Henry (1915) and taken up by numerous other authors but not collected for over 75 years.

***Bromus macounii* Vasey** = *B. erectus*

***Bromus orcuttianus* Vasey** First cited by Macoun (1888) and taken up by Henry (1915) and Hubbard (1969) but probably based on a misidentification since the species is known north only to Mt. Rainer, WA.

***Bromus rubens* L.** Cited from Nanaimo by Henry (1915) and taken up by Taylor and MacBryde (1977) and Scoggan (1978) but not collected for over 75 years.

***Bromus suksdorfii* Vasey** Taylor and MacBryde (1977) and Straley *et al.* (1985) reported this species from Manning Park but it is based on a misidentification of *B. vulgaris*.

***Corynephorus canescens* (L.) Beauv.** This species was once established in Vancouver (Eastham 1947, Hubbard 1969, Taylor and MacBryde 1977) but no longer persists.

***Danthonia parryi* Scribn.** First reported by Baum and Findlay (1973) and taken up by Taylor and MacBryde (1977) but no material seen during this study.

***Danthonia sericea* Nutt.** This species, cited by Baum and Findlay (1973), occurs east of the Mississippi River in the US.

***Echinochloa colonum* (L.) Link.** Cited by Hubbard (1969) but no specimens seen.

***Elymus arenarius* L.** (*Leymus arenarius* [L.] Hochst.) This species, reported by Taylor and MacBryde (1977) and Scoggan (1978), does not occur in BC and probably represents a misidentification of *L. mollis.*

***Elymus scribneri* (Vasey) M.E. Jones** Reported by Hitchcock *et al.* (1969) for BC but no material found.

***Elymus virginicus* L. var. *submuticus* Hook.** Cited by Taylor and MacBryde (1977) and Scoggan (1978), this species does not occur W of the Rocky Mountains.

***Festuca cinerea* Vill.** Reported from only one location in BC, near Nanaimo Lakes on Vancouver Island in 1950.

***Festuca hallii* (Vasey) Piper** Cited by Taylor and MacBryde (1977) this species does not occur in BC.

Festuca ovina* L. var. *ovina This species, cited by Scoggan (1978), does not occur in BC.

***Festuca vivipara* (L.) Smith** (*Festuca ovina* L. var. *vivipara* L.) Reported by Taylor and MacBryde (1977), this species does not occur in BC.

***Glyceria declinata* Brebiss.** No material of this European introduction, reported by BC for Scoggan (1978), was found during this study.

***Glyceria fluitans* (L.) R. Br.** Cited by Boivin (1966–1967) and Scoggan (1978) but no BC material seen.

***Glyceria leptostachya* Buckl.** This species, listed by Eastham (1947), Hubbard (1969), Taylor and MacBryde (1977) and Scoggan (1978), was based on a single collection made at Alberni over 75 years ago.

***Melica aristata* Thurb. ex. Boland.** This species of the W U.S.A. was first reported by Henry (1915) and later taken up by Hubbard (1955) and Taylor and MacBryde (1977). These reports are apparently erroneous.

***Melica geyeri* Munro** This report by Eastham (1947) was based on an early (1917) Vancouver Island collection.

***Mibora minima* (L.) Desv.** Known only from an old (1914) collection at the Experimental Station near Sidney.

***Muhlenbergia sylvatica* Torr.** Collected once from a roadside near Kamloops (Scoggan 1978) but never recollected.

***Muhlenbergia uniflora* (Muhl.) Fern.** This species was probably introduced from ship ballast and has not be collected since 1909.

***Panicum rigidulum* Bosc ex Nees** (*Panicum agrostoides* Spreng. *sensu* Hubbard [1969]) Cited from Sproat Lake by Macoun (1888) and more recent authors but never recollected in BC.

***Phalaris minor* Retz.** Collected on ballast at Nanaimo about 100 years ago and cited by Taylor and MacBryde (1977) and Scoggan (1978) but never recollected in BC.

***Phippsia algida* (Soland.) R. Br.** This species, cited by Taylor and MacBryde, has yet to be collected in BC. It occurs just north of the BC border in SW YT.

Poa abbreviata* R. Br. ssp. *abbreviata Collections treated as this high arctic species by Taylor and MacBryde (1977) have been reidentified as *P. abbreviata* ssp. *pattersonii.*

***Poa arctica* R. Br. ssp. *grayana* (Vasey) Löve, Löve & Kapoor.** (*Poa grayana* Vasey) This species of the Rocky Mountains, reported by Hitchcock *et al.* (1969) and Taylor and MacBryde (1977), does not reach Canada. Material given this name has been placed under other subspecies of *P. arctica.*

***Poa arida* Vasey** Reports of this species (Scoggan 1978) were based on collections of *P. secunda.*

***Poa bolanderi* Vasey** Collections of this species, cited in Henry (1915), have been re-identified as *P. howellii.*

Poa cusickii* Vasey ssp. *cusickii This species of the W US does not reach our region. Material identified as this species, and reported by Taylor and MacBryde (1977) and Scoggan (1978), has been re-identified as *P. cusickii* ssp. *pallida*.

***Poa kelloggii* Vasey** First reported by Jamieson (1979) and taken up by Straley *et al.* (1985), this species is apparently restricted to N CA. The specimen in question has been re-identified as *Poa marcida.*

***Poa leibergii* Scribn.** (*P. vaseyochloa* Scribn.) Cited by Henry (1915) but no material seen.

***Poa occidentalis* Vasey.** Reported by Henry (1915) and Hubbard (1955) and taken up by Scoggan (1978). The relevant material has been re-determined as *P. palustris.*

***Poa reflexa* Vasey & Scribn.** Cited by Hitchcock *et al.* (1969) and Taylor and MacBryde (1977) but our material is actually *P. leptocoma.*

***Poa suksdorfii* (Beal) Vasey** The relevant collection cited by Douglas and Ratcliffe (1981) has been re-identified as *P. secunda* ssp. *secunda.*

***Poa tracyi* Vasey** First cited by Hitchcock (1951) and taken up by Taylor and MacBryde (1977), this species occurs only in the S Rocky Mountains.

***Puccinellia lemmonii* (Vasey) Scribn.** Most collections from BC identified as such, and reported by Taylor and MacBryde (1977) and Scoggan (1978), are depauperate specimens of *P. nuttalliana.*

***Puccinellia maritima* (Huds.) Parl.** This European species is known only from N American ports (presumably on ballast). Most collections from BC given this name are actually robust specimens of *P. nutkaensis* (i.e., *P. lucida* Fern. & Weath).

***Spartina pectinata* Link.** Cited by Eastham (1947), Taylor and MacBryde (1977) and Scoggan (1978) from a collection made over 40 years ago at Sea Island.

***Stipa occidentalis* Thurb. ex S. Wats.** This taxon, cited for BC by Hitchcock *et al.* (1969) and Taylor and MacBryde (1977), does not occur in BC.

***Zizania aquatica* L.** Cited by Taylor and MacBryde (1977) but not collected for over 40 years.

PONTEDERIACEAE

***Pontederia cordata* L.** This species, cited by Taylor and MacBryde (1977) and Scoggan (1978), was based on a Port Alberni collection made in 1950. A more recent collection (Brayshaw 1985) was taken from a waterfront planting outside a home in the Victoria area. The plant does not spread in our region.

POTOMOGETONACEAE

***Potamogeton diversifolius* Raf.** Reported by Hitchcock *et al.* (1969) and Brayshaw (1985), but no specimen seen. No BC specimen cited in Reznicek & Bobbette (1976).

***Potamogeton fibrillosus* Fern.** The specimen (collected in Manning Park) on which a report of this species was based is poor (with immature fruits) and difficult to identify. Haynes & Reveal (1973) and Haynes (1974) treated this taxon as *P. foliosus* var. *fibrillosus* (Fern.) Haynes & Reveal, characterized by fibrous stipules.

SPARGANIACEAE

Sparganium americanum **Nutt.** Although treated as a BC species by various authors (e.g., Fernald 1950, Boivin 1966–1967, Taylor and MacBryde 1977, Brayshaw 1985), no material was located during this study. It is likely that the material referred to by Fernald (1950) and others actually belongs to the closely related *S. angustifolium* complex.

GLOSSARY

Achene – A dry, single-seeded fruit that does not open at maturity.

Acicular – Shaped like the needle of a pine tree.

Acorn – A one-seeded fruit of oak trees.

Acuminate – Tapering to a narrow tip or concave point, the sides generally concavely narrowing.

Acute – Gradually tapering to a point, the sides straight (i.e., not convex).

Adaxial – On the side of a structure, turned toward the main axis.

Adnate – Fusion or attachment of unlike parts (e.g., the calyx and corolla).

Adventitious – Development of structures in unusual locations; in the case of roots, generally refers to the development of roots from the stem or leaf.

Alternate – Any arrangement of leaves or other parts not opposite or whorled; arranged singly at different heights on the stem.

Ament – A pendulous spike-like inflorescence with apetalous, unisexual flowers; a catkin.

Amphiberingian – Occurring on both sides of the Bering Straits; a phytogeographic element resulting from past migration across the Bering Land Bridge.

Amplexicaule – Clasping the stem.

Androecium – A collective term referring to male floral parts, that is, the stamens.

Androgynous – A spike bearing staminate (male) and pistillate (female) flowers where the staminate flowers are positioned above the pistillate flowers, closer to the apex of the inflorescence.

Annual – Living for only one growing season.

Anthesis – The period of time when a flower is fully open.

Apex – Tip, point or end.

Aphyllopodic – Having lower leaves that are bladeless or reduced to scales (e.g., *Carex* spp.).

Apical – Located at the apex.

Apomictic – Capable of reproducing asexually.

Apomixis – Any of several types of asexual reproduction.

Aquatic – Living in water.

Arachnoid – Covered with long, delicate, tangled hairs; resembling a cobweb.

Aristate – Having an awn or a bristle at the tip.

Armed – Possessing prickles, spines, or thorns.

Articulate – Jointed; separating at a joint, by a clean cut scar, upon maturity.

Anther – The pollen producing sac of the stamen.

Attenuate – Gradually tapering to a slender base or apex.

Auct. non – Indicates that a species name is used commonly in the literature but is incorrect and is not the name used to refer to the type specimen.

Auricle – An ear-shaped lobe or appendage.

Auriculate – With auricles; often used to describe the base of a leaf blade where there is an auricle on either side of the petiole.

Awn – A slender bristle-like appendage, usually at the apex of a structure.

Axil – The angle between a structure and the axis to which it is attached (e.g., the angle between a leaf and the stem).

Axile placentation – A pattern of placentation where the ovules are attached to the partitions of a compound pistil near the center of the ovary.

Axillary – Used in reference to structures attached in the axil.

Banner – The largest upper petal of a papilonaceous flower.

Barb – A short, firm, reflexed process.

Barbellate – Minutely or finely barbed; bearing stiff hairs with barbs.

Basal – At, or forming, the base.

Beak – A long, slender projection on a broader structure such as a fruit or seed.

Bi – A prefix meaning two or twice.

Biennial – Living for two seasons.

Bilabiate – Two-lipped; often with reference to a zygomorphic flower.

Bipinnate – *See* bipinnatifid.

Bipinnatifid – Twice branched.

Biserrate – Doubly serrated.

Bisexual – Bearing both male and female reproductive structures.

Biternate – Two groups of three.

Bloom – *See* glaucous.

Bractlet – A small or secondary bract (syn. bracteole).

Bract – A modified leaf, reduced in size or form that is associated with a flower or an inflorescence; in conifers, a main appendage arising from the axis of a cone.

Bristle – A stiff, hair-like structure.

Bristly – With bristles.

Bulb – A small, stout underground stem bearing modified (scale-like) leaves and fibrous basal roots; generally food storage organs.

Bulbiferous – Producing bulbs.

Bulbil – A small bulb produced above the ground, often on the flowering stem.

Caducous – Falling off the plant early relative to other similar structures.

Caespitose – Growing in dense, low-lying clumps.

Calcareous – Rich in calcium carbonate.

Callose – Bearing callosities.

Callosity – *See* callus.

Callus – A hard or firm thickening of a structure; in grasses, the thick, hardened basal portion of the lemma.

Calyx – A collective name referring to the sepals, the outer whorl of the perianth.

Campanulate – Bell-shaped to saucer-shaped.

Cane – In *Rubus* spp., lignified shoots that arise each year.

Canescent – Appearing white or gray in colour due to very short, fine hairs.

Capillary – Very slender and hair-like.

Capitate – In a head-like cluster.

Capsule – A dry, dehiscent fruit comprised of greater than one carpel; method of splitting further classifies the capsule type (e.g., circumscissile, septicidal, etc.).

Carpel – The basic unit of a pistil, a modified seed bearing leaf.

Caruncle – An often fleshy appendage at or near the attachment point (hilium) of some seeds.

Caryopsis – The dry, one-seeded fruit found in grasses.

Caudate – Bearing a terminal, long and slender appendage like a tail.

Caudex – The woody, thickened and persistent basal portion of a herbaceous perennial that gives rise to leaves and flowering stems each year.

Caulescent – Having an obvious leafy stem.

Cauline – Of, or pertaining to, the stem.

Cellular – A term used in reference to the cell-shaped network of minutely raised and lignified ridges on epidermis (e.g., on the culm of some *Carex* spp.).

Chaffy – With thin dry scales or bracts.

Chartaceous – Having the texture of stiff paper or parchment; membranous.

Ciliate – Having a fringe of fine hairs (cilia) along the margin of a leaf.

Ciliolate – Ciliate but with smaller hairs.

Cinereous – Appearing grey in colour because of short hairs.

Circumboreal – Distributed at the top or the bottom of the world in the boreal zone.

Circumpolar – Distributed around the North or the South Pole.

Circumscissile – A capsule that dehisces transversely in a circular line, like the opening of a lid.

Clavate – Club-shaped, the wider portion at the apex.

Claw – The narrowed, basal portion of some petals or sepals.

Cleft – Usually applied to leaf margins and sinuses, cut sharply and about half-way to the midrib or base; deeply lobed.

Collar – In grasses, the band on the outside area of the leaf at the point where the leaf blade and the sheath converge.

Coma – A tuft of hair usually at the apex of structures like seeds.

Compound – With two or more parts or components.

Confluent – Structures merging into one; fusing at the bases.

Conic – Cone-shaped.

Connate – Union or attachment of like parts (e.g., petals).

Connivent – Convergent but not fused.

Convolute – The arrangement of petals or sepals such that each petal or sepal has one edge exposed while the other is covered; in grasses, a leaf that is rolled longitudinally so that one edge is inside the other.

Cordate – Shaped like a heart, with the notch at the base.

Coriaceous – Having a thick and leathery texture.

Corm – A thickened, underground stem with papery leaves; a storage organ.

Corniculate – Bearing small horns.

Corolla – The collective term describing the petals of a flower.

Corymb – A flat-topped, indeterminant inflorescence where pedicels of florets are progressively shorter towards the center of the inflorescence; may be simple or compound.

Corymbose – A flat or convex-topped inflorescence; resembling a corymb.

Costa – A prominent rib or vein.

Cotyledon – The first functional leaf of a seedling.

Crenate – Toothed with rounded, broad teeth.

Crenulate – Crenate but with smaller teeth.

Crown – The persistent basal portion of a perrenial plant (e.g., grass), or the leafy portion of a tree; any appendage situated between the corolla and the stamens.

Cruciform – Having the shape of a cross.

Cucullate – Hooded or hood-shaped.

Culm – The flowering stem of grasses, sedges and rushes.

Cuneate – Triangular or wedge-shaped with the acute angle at attachment point.

Cupulate – Cup-shaped.

Cuspidate – Bearing a short, rigid and sharp point, usually in reference to the apex of a leaf.

Cyme – An inflorescence in which all floral axes terminate in a single flower with the terminal flower, at the apex of the inflorescence, blooming first.

Cymose – Bearing cymes, or cyme-like.

Deciduous – Shedding of leaves seasonally or structures after the growth period.

Decumbent – Lying on the ground with the end curved upward.

Decurrent – A structure adnate or confluent to the stem near its insertion point.

Definite – Distinct; precise.

Deflexed – Bent downward or backward.

Dehiscent – Discharge of contents by the opening of pores or slits, most often pertaining to the opening of fruit at maturity.

Deltoid – Shaped like an equilateral triangle.

Dentate – With sharp, spreading teeth.

Dichotomous – Forked with both branches of approximately equal size.

Didymous – Occurring or developing in pairs.

Dimorphic – Having two forms.

Dioecious – Male and female reproductive structures borne on different plants.

Disarticulate – The separation of structures at maturity.

Disciform – Circular and like a disk.

Discoid – Like a disk; in the Asteraceae, refers to a flower head where all flowers are disk flowers.

Disk corolla (disk flower) – Tubular corolla or flower of some members of the Asteraceae.

Dissected – Divided into many small, fine and often repeated parts or segments.

Dorsal – Pertaining to, or located on, the back of an organ.

Drupe – A fleshy, one-seeded or one-stoned fruit of some plants of the rose family (e.g., cherries, peaches).

Elliptic – Shaped like an ellipse, the broadest point midway between the ends, and the width about one-half the length.

Emarginate – Having a shallow notch at the apex, usually with reference to a leaf.

Endocarp – The innermost layer of the three layers comprising the mature ovary wall of a fruit.

Entire – A continuous margin, without teeth or divisions.

Epigaeous – Growing above ground.

Epiphytic – A plant that grows on another plant but does not receive nourishment from it.

Equitant – With alternate leaves whose basal portions overlap and are flattened lengthwise (e.g., Iris).

Erose – An irregular margin, appearing eroded or gnawed, not fringed or toothed.

Ex – With respect to plant authorities, indicates that the first author proposed the name for a plant and the second author published that name.

Exserted – Projecting beyond a surrounding structure, like stamens projecting beyond a corolla.

Extravaginal – Located outside the sheath that surrounds another organ.

Falcate – Sickle-shaped; curved sideways and tapering.

Farinose – Covered with a white mealy substance.

Fascicle – A cluster.

Ferruginous – Rust in colour.

Filament – The slender stalk of the stamen that supports the anther.

Filiform – Thread-like, slender and cylindrical.

Fimbriate – Fringed with hairs coarser than cilia.

Fistulose – Hollow.

Flabellate – Fan-shaped.

Flabelliform – *See* flabellate.

Flexuous – Bent in opposite directions; having a more-or-less zig-zag or wavy form.

Floccose – Covered with tufts of soft woolly hairs that usually rub off readily.

Floret – A small flower, particularly one in a floral cluster; pertains also to the flower of a grass spikelet which includes the lemma, palea and reproductive structures.

Foliaceous – Leaf-like.

Follicle – A dry, dehiscent fruit comprised of one carpel that splits at the side upon maturing to release its seeds.

Fornix (pl. fornices) – A small, scale-like or petaloid appendage in the tube or throat of the corolla, found in many borages.

Free central placentation – A pattern of placentation where the ovules are borne on a single stalk, anchored at the base of a one-celled ovary.

Frond – The leaf of a fern.

Fruit – The ripened ovary and any other structures that enclose it.

Funnelform – Having the shape of a funnel, usually pertaining to the corolla.

Fusiform – Spindle-shaped, broadest at the middle and tapering to opposite ends.

Galea – The hooded portion of the perianth in some irregular or bilabiate flowers.

Gametophyte – The sexual reproductive stage of a plant that produces gametes, in ferns it is often the small inconspicuous form of the plant.

Gemma (pl. gemmae) – A young bud from which plants vegetatively reproduce.

Geniculate – Abruptly bent.

Gibbous – Swollen on one side.

Glabrate – Nearly or becoming glabrous, usually with age.

Glabrescent – *See* glabrate.

Glabrous – Smooth, without hairs of any type.

Gland – A spot or structure that produces a sticky or greasy substance.

Glandular – Having glands.

Glandular-pubescent – With gland-tipped hairs.

Glaucous – Covered with a whitish or bluish waxy powder that readily rubs off.

Globose – Spherical.

Glochidia – Retorsely barbed hairs.

Glomerate – Condensed in heads.

Glomerule – A dense cluster, usually of flowers.

Glumes – The two small bracts located at the base of a grass spikelet, they do not subtend flowers.

Glutinous – Covered with a sticky, glue-like resin.

Gynaecandrous – A spike bearing staminate (male) and pistillate (female) flowers where the staminate flowers are positioned below the pistillate flowers, thus the pistillate flowers are closest to the spike apex.

Hastate – Shaped like the head of an arrow with the basal lobes flaring outward.

Head – A dense indeterminant inflorescence of sessile (or nearly sessile) flowers, situated on an axis that is much compressed vertically and sometimes flattened or saucer-shaped.

Hemispheric – In reference to a head, with a half-sphere shape.

Herb – Plant with stem that dies back to the ground after flowering or at the end of the growing season.

Herbaceous – A non-woody plant.

Heterophyllus – Having two types of leaves.

Heterosporous – Production of two different types of spores, generally differing in size.

Hirsute – Covered with long, stiff and coarse hairs.

Hirtellous – Minutely hirsute.

Hispid – Bearing long, rigid and very sharp bristles or bristle-like hairs.

Hispidulous – Slightly hispid.

Hoary – Covered with short and fine white or gray hairs.

Homophyllus – With leaves that are all identical or very similar in form.

Homosporous – Production of one type of spore.

Hyaline – Thin, almost completely transparent tissue.

Hybrid – A cross between two taxa, usually a cross between two species of the same genus.

Hypanthium – A cup-like structure surrounding the ovary, formed by the fusion of the basal portions of the perianth, calyx, corolla and androecium.

Imbricate – Partly overlapping in a spiral manner, shingle-like.

Imperfect – Flowers lacking either female or male reproductive structures, a unisexual flower.

In – With respect to plant authorities, indicates that the first author described the plant in an article edited or published by the second author.

Incised – Cut sharply, deeply and irregularly, as in leaf margins.

Indument – Appendages or outgrowths from the epidermis of a plant.

Indurate – Hard or stiff.

Indusium (pl. indusia) – A scale-like outgrowth that covers and protects the cluster of sporangia located on the underside of a fern frond.

Inferior (ovary) – When the perianth and androecium are fused at the top of the ovary wall, the ovary position is described as inferior.

Inflorescence – The flowering cluster, categorized by the arrangement of flowers on the floral axis.

Intergradients – Taxa that interbreed and consequently share many characteristics making it difficult to separate them taxonomically.

Internode – The section of a stem or another structure between two nodes.

Interrupted – Discontinuous.

Involucre – A rosette of bracts subtending an inflorescence.

Involute – Margins longitudinally rolled in, on the upper side, toward the center.

Irregular – Pertaining to a flower in which similar parts (e.g., petals) are unequal in shape and size (syn. zygomorphic).

Keel – In a papilonaceous flower, the two lowermost connate petals; resembles the keel of a boat; in grasses, the sharp crease or ridge of the leaf blades, the lemma or the glume.

Lacerate – With an irregular or jagged margin, as if torn.

Laciniate – Cut into narrow, often pointed, lobes.

Lamina – The flattened portion (i.e., the blade) of a leaf or petal.

Lanate – Woolly, with long intertwined, curly hairs.

Lanceolate – Lance-shaped; longer than broad, widening above the base and tapering to the apex.

Leaflet – One of the segments or divisions of a compound leaf.

Lemma – The outermost of the two bracts that subtend the grass floret (*See* palea).

Lenticular – Biconvex in shape.

Ligulate – With a ligule.

Ligule – Flattened part of the ray corolla in the Asteraceae; in many grasses and some sedges and rushes, a membranous (or hair-like) appendage on the inner side of the leaf at the junction of the leaf blade and the sheath.

Linear – Long and narrow, with parallel margins.

Locule – A compartment in the ovule or anther.

Lustrous – Shiny.

Lyrate – Pinnatifid with the end lobe enlarged and rounded.

Malpighiaceous – Hairs attached to a surface by their centers.

Maritime – Near to, or influenced by, the sea.

Megasporangia – The sporangia of a heterosporous plant that produces megaspores.

Megaspore – On a heterosporous plant, the larger of two spore sizes produced; a spore involved in development of female reproductive organs.

Mericarp – One portion of the fruit (e.g., a schizocarp) that splits into two halves at maturity, found in members of the Apiaceae.

Merous – A suffix referring to the division of floral parts, e.g., a 3-merous flower would have 3 petals, 3 sepals, 3 stamens, and a pistil with 3 carpels.

Mesic – Relating to habitats with a moderate amount of moisture.

Microsporangia – The sporangia of a heterosporous plant that produces microspores.

Microspores – On a heterosporous plant, the smaller of two spore sizes produced; a spore involved in the development of male reproductive organs.

Monoecious – Male and female reproductive structures borne on the same plant.

Mottled – Marked with spots.

Mucro – A small, short and abrupt point.

Mucronate – Having mucro.

Muricate – Roughened with short, firm and sharp projections.

Naked – Lacking various organs, hairs or appendages.

Naturalized – A well established plant originating from another area.

Neutral – In reference to a flower lacking functional stamens and pistils.

Nodding – Curving to one side; minutely pendulous.

Node – The location on a stem where stems or leaves usually originate, generally a swollen area, other structures like tendrils and flowers may also originate here.

Nom. superfl. – A plant name that has been superceded by a valid name; a superfluous name.

Nut – A hard, indehiscent fruit, with one seed.

Nutlet – A small nut; also used when referring to each lobe of the 4-lobed nuts found in borages.

Ob – A prefix meaning in the reverse of the typical direction.

Obconical – Conical in shape with attachment point at the narrow end of the cone.

Obcordate – Having an inverted heart shape.

Oblanceolate – Reverse of lanceolate, widest above the middle and tapering to the base.

Oblong – Much longer than wide with parallel or nearly parallel sides.

Obovate – The reverse of ovate; attached at the narrow end.

Obpyramidal – Having the shape of an inverted pyramid.

Obsolete – Reduced to the point of being entirely, or almost, undetectable.

Obtuse – Blunt, rounded.

Ochroleucous – Yellowish-white.

Olivaceous – Olive-like in form or colour.

Orbicular – Having a circular outline.

Ovary – The enlarged base of the pistil that contains the developing seeds, matures into the fruit.

Ovate – Shaped like a hen's egg, attached at the broader end.

Palea – The innermost of the two bracts that subtend the floret in grasses (*See* lemma).

Palmate – Lobed or divided in a palm- or hand-like fashion, digitate.

Panicle – A branching inflorescence on which younger flowers are borne at the apex.

Paniculiform – Resembling a panicle.

Papillate – Covered with small minute rounded protuberances (papillae).

Papilonaceous flower – A strongly irregular flower with five petals, the lowermost two petals are connate and form the keel.

Pappus – The modified outer perianth series crowning the ovary and achene in the Asteraceae; usually plumose, bristle-like, or of scales.

Parietal placentation – A pattern of placentation, in compound pistils only, where ovules are attached to the outer wall of the carpel.

Pectinate – Comb-like, pinnately lobed with narrow segments arranged like the teeth of a comb.

Pedicel – The stalk of a single flower in an inflorescence, or of a spikelet in grasses.

Pedicellate – Having a pedicel.

Peduncle – The single supporting stalk of a flower or inflorescence.

Peltate – An umbrella-shaped leaf whose attachment point is on the underside of the leaf rather than at the base or margin.

Pendulous – Hanging; resembling a pendulum.

Perennial – Growing for more than two seasons.

Perfect – A flower that has functional stamens and pistils.

Perfoliate – Bearing opposite leaves whose bases are wrapped around the stem and connate, giving the appearance that the stem passes through the leaf.

Perianth – A collective term for the calyx and corolla.

Perigynium – A saccate bract surrounding the pistil or achene in *Carex* spp.

Persistent – Remaining attached.

Petiole – Leaf stalk.

Phenotypic plasticity – Ability to adapt, morphologically, to changing environmental conditions.

Phyllode – A somewhat expanded but bladeless petiole taking on the function of a leaf.

Phyllopodic – Having well developed lower leaves instead of scales.

Piliform – Having the shape of a long, sometimes wavy, hair.

Pilose – Shaggy with soft hairs.

Pinna (pl. pinnae) – One of the primary divisions of a pinnately compound leaf or frond.

Pinnate – Compound leaf with the leaflets on both sides of an elongated axis.

Pinnatifid – Pinnately lobed.

Pinnule – A secondary leaflet in a bipinnately compound leaf.

Pistillate – Flowers possessing pistils, but no functional stamens (i.e., a female flower).

Pith – The spongy tissue in the center of a root or stem that surrounds vascular bundles.

Placentation – The pattern of attachment of ovules in the ovary. *See* axile, parietal or free central placentation.

Plumose – Feather-like.

Pod – Any type of dry, indehiscent fruit.

Pollen – A general term pertaining to the haploid (n) microspores produced in the anther.

Polypetalous – Flowers with all petals distinct, not united.

Polyploidy – Two or more sets of chromosomes (e.g., 3n, 4n, 8n).

Prehensile – Grasping.

Prickle – Small, sharp outgrowth from epidermis, contains no vascular tissue.

Prismatic – Prism-shaped.

Procumbent – Prostrate or trailing stem, does not usually root at nodes.

Propagule – A newly growing structure.

Pro parte – In part.

Pro parte sed non typus – Resembles, in part, but not completely the type specimen; possessing some but not all characteristics of the type specimen.

Prostrate – Lying flat on the ground, may root at the nodes.

Pruinose – *See* glaucous.

Puberulent – Bearing very short, erect hairs, scarcely visible to the unaided eye, minutely pubescent.

Pulverulent – Dusty, powdery.

Pulvinus – Swollen tissue at the insertion point of the leaf petiole.

Punctate – Marked with pits or dots formed by glands or waxy spots.

Puncticulate – Minutely punctate.

Pyriform – Pear-shaped.

Raceme – An indeterminant, elongated inflorescence where single flowers are borne on pedicels arranged on a single axis; the youngest flowers are at the apex.

Racemiform – Possessing the form of a raceme.

Rachilla – The axis of a spikelet in grasses, refers to a secondary axis in sedges.

Rachis – The main stalk of an inflorescence or compound leaf.

Radiate – In the Asteraceae, having ligulate or ray flowers.

Ray (flower) – Strap-shaped corolla (ligule).

Receptacle – The expanded end of the stem on which floral parts are borne.

Reclining – Lying on the ground or one plant pressed on another.

Reflexed – Bent sharply downward or backward.

Regular (flowers) – A flower with radial symmetry; petals and sepals that are similar in shape, size, colour and orientation.

Reniform – Kidney-shaped.

Resinous – Covered with a resin.

Reticulate – Forming a network, netted like the veins of a leaf.

Retorse – Directed backward or downward.

Retuse – *See* emarginate.

Revolute – Leaf margins rolled toward the underside of the leaf.

Rhizome – A creeping, underground stem or rootstalk that roots at the nodes.

Rhombic – Diamond-shaped.

Rib – A prominent vein, usually on a leaf.

Root – The descending axis of a plant that anchors and absorbs moisture from the ground, lacks nodes and internodes.

Rootstock – Used loosely to mean a main root or an underground root-like stem or branch.

Rosette – A dense cluster of flowers or leaves radiating around a central axis, usually basal in the case of leaves.

Rotate – A saucer-shaped corolla where the corolla tube is short and petals are generally reflexed.

Rotund – Rounded.

Rudimentary – Usually a small, imperfectly developed structure, generally reminiscent of an earlier evolutionary stage.

Rufous – Reddish-brown.

Rugose – Possessing a wrinkled surface.

Rugulose – Slightly rugose.

Saccate – Shaped like a sac or pouch.

Sagittate – Shaped like the head of an arrow, triangular with the basal lobes pointing backward.

Salverform – A corolla having a long slender tube that flares abruptly at the end.

Samara – A winged, indehiscent fruit.

Samaroid – Resembling a samara.

Saprophytic – A plant that derives its nutrients from dead organic matter and lacks the chlorophyll that gives most plants their green colour.

Scaberulous – Slightly scabrous.

Scabrid – *See* scabrous.

Scabridulous – *See* scaberulous.

Scabrous – Having a roughened surface caused by very short stiff hairs or short sharp projections.

Scabrule – A stiff pointed projection or a short stiff hair.

Scale – A small, flat structure resembling the scale of a fish.

Scape – A flowering stem without leaves.

Scapose – Bearing a scape.

Scarious – Having chaffed, membranous or brown edges.

Scrambling – Slightly reclining.

Scree – Broken rock material characteristic of alpine areas, usually a result of freeze-thaw cycles.

Secund – Having all flowers or branches borne on one side of the axis.

Semi-aquatic – A plant that may tolerate partial submersion in water or a plant that roots in the soil but produces aquatic leaves.

Sensu lato – In the broad sense.

Sepals – The outermost whorl of floral leaves, also called the calyx and is usually green in colour.

Seriate – In a series of rows.

Sericeous – A surface with long, soft, appressed hairs, giving a silky appearence and texture.

Serrulate – A leaf margin with small serrated teeth.

Sessile – Without a stalk.

Seta (pl. setae) – A bristle.

Setose – Bearing setae.

Sheath – An thin tubular organ that partially or wholly surrounds another organ; in grasses, the basal portion of the leaf blade that wraps around the stem.

Shrub – A woody perennial plant, less than 10 m tall, with several basal stems.

Silicle – A dry, indehiscent fruit of plants in the Brassicaceae generally not more than two times longer than wide.

Silique – A long slender, indehiscent fruit of the Brassicaceae family.

Simple – A leaf not compounded into leaflets or an unbranched inflorescence.

Sinuate – A pronouncedly waved leaf margin.

Sinus – A cavity or a depression between two adjacent structures (e.g., the space between two lobes of a leaf).

Sordid – Dirty tinge.

Sorus (pl. sori) – A cluster of spore producing structures (sporangia) on the underside of a fern frond.

Spadix – A spike with small, densely crowded, unisexual flowers on a thick and fleshy axis, the inflorescence is subtended by a spathe.

Spathe – A large bract surrounding a spadix.

Spatulate – Broad and rounded at the apex; narrowing to the base.

Spiciform – An inflorescence resembling a spike.

Spike – An unbranched inflorescence; flowers are sessile.

Spikelet – A small spike; in grasses, the primary inflorescence composed of two glumes and one or more florets.

Sporangia – Spore producing structures.

Sporangium – Reproductive structure of ferns that produces spores.

Spore – A haploid (n) reproductive cell.

Sporocarp – The structure that contains the sporangia in ferns.

Sporophore – A leaf bearing reproductive structures; generally used in reference to the spore bearing leaves of ferns.

Sporophyll – A modified leaf bearing sporangia.

Sporophyte – The generation in the reproductive cycle of plants that produces spores; in ferns generally refers to the conspicuous form of the plant (*See* gametophyte).

Spur – A slender and hollow floral appendage, generally nectariferous.

Squamiform – Scale-like.

Squarrose – Having parts, generally at the tip of a structure, that are abruptly curving and spreading.

Stamen – The male sex organ of a plant constructed of an anther and a filament, the anther produces pollen.

Staminate – Flowers bearing stamens but no functional pistils; a male flower.

Staminode (pl. staminodia) – A sterile, sometimes modified stamen (syn. staminodium)

Stellae – Star-shaped structures

Stellate – Star-shaped; radiating out from a common center.

Steppe – An extensive area of dry, arid grassland.

Sterile – Infertile flowers.

Stigma – The generally expanded, apical portion of the pistil which is receptive to pollen.

Stipe – A stalk supporting the pistil; also the petiole of the frond of a fern.

Stipitate – Borne on a stipe or short stalk.

Stipule – An appendage on each side of the base of a leaf or petiole at its insertion point on the stem.

Stolon – A reclining above-ground horizontal stem that roots at the nodes.

Stramineous – Having the colour of straw.

Strigillose – Like strigose but with shorter hairs.

Strigose – A surface having sharp, appressed and straight hairs that are stiff and often basally swollen.

Style – A generally elongated stalk connecting the ovary and the stigma through which the pollen tube grows.

Styliform – Resembling a long slender stalk, like a style.

Stylopodium – A disk-like enlargement at the base of the style.

Sub – A prefix signifying below, somewhat, slightly or rather.

Subspecies (ssp.) – A category of biological classification ranking immediately below the species level.

Subtend – To be positioned directly or closely beneath something.

Subulate – Narrowly triangular and tapering to a small point like an awl.

Succulent – Fleshy, having a soft and thickened texture.

Sulcate – Longitudinally grooved or furrowed.

Sympatric – Occurring in the same area.

Sympodial – Having an apparent main or central axis that is actually comprised of several short overlapping branches.

Synonym – An equivalent superceded name, a second name to a given taxon.

Talus – An accumulation of rock material on or at the base of a slope.

Taproot – An underground tapering root from which smaller lateral roots may form (may be thick or thin).

Tendril – A long, slender, generally twisting structure from which a climbing plant acquires its support.

Tepals – A collective term applied to petals and sepals when they are morphologically identical.

Terete – A structure that is circular in cross section but imperfectly cylindrical since the object may taper one or both ways.

Ternate – Arranged in groups of three.

Tetrad – In groups of four.

Thyrse – An elongated and densely flowered panicle, often contracted and pyramidal in shape.

Thyrsoid – Resembling a thyrse.

Tomentose – Densely pubescent with soft, matted wool-like hairs.

Tomentulose – Slightly tomentose.

Toothed – Bearing teeth-like protrusions.

Torose – Having successive swellings and contractions.

Torulose – Slightly torose, like the small constrictions of a pod between seeds.

Tri – A prefix meaning three or three times.

Tricuspidate – *See* cuspidate.

Trifid – Separation into three parts.

Trifoliate – A compound leaf with three leaflets.

Trigonous – Having three angles.

Tripinnate – Branched three times.

Triquetrous – With three sharp angles, like a triangle with concave sides.

Trophophore – A vegetative leaf or frond, one that does not have reproductive structures attached.

Truncate – Cut-off at the end; having a squared shape at the apex or base.

Tuber – A thickened, modified underground stem having many buds (e.g., a potato).

Tubercle – A small, rounded swelling.

Tuberculate – Bearing tubercles.

Tuberous – Thickened; producing tubers.

Turbinate – Inversely conical, top-shaped.

Turgid – Swollen.

Turion – A small, scaly bud-like offset on an underground stem.

Type specimen – A specimen identified by the author of a species that, morphologically, represents the best example of this species (syn. holotype).

Umbel – An inflorescence in which all flowers arise from a common point, younger flowers are in the center, may be simple or compound.

Umbellate – Arranged in umbels.

Umbellet – A secondary umbel in a compound umbel.

Undulate – A slightly wavy margin.

Uniseriate – Arranged in a row or series.

Valve – One of the segments into which a dehiscent fruit splits at maturity.

Variety (var.) – A category of biological classification ranking immediately below the species or subspecies level.

Veins – Thin threads of vascular tissue, as in a leaf.

Velutinous – Having a velvety texture.

Vernal – Pertaining to, or appearing in, spring.

Verrucose – Covered with wart-like protrusions.

Verticillate – *See* whorled.

Verticils – *See* whorl.

Vestige – The reduction of a structure whose functional importance has decreased with evolutionary time.

Villous – With long and soft, unmatted hairs.

Viscid – Glutinous or greasy.

Wanting – Lacking.

Whorl – Three or more structures arranged in a circular manner around a common insertion point.

Whorled – Having whorls.

Wing – A thin membranous outgrowth.

Winged – Bearing wings.

INDEX

Note: This index is a compilation of the indices from Parts 1 through 4 of the Vascular Plants series. Taxa from Parts 1 through 3 are referenced with the volume number followed by its page number (e.g., ***Abies*** 1–5; buttercup 3–24). Entries from this volume (Part 4) are followed by a page number only.

Note: This index is a compilation of the indices from Parts 1 through 4 of the Vascular Plants series. Taxa from Parts 1 through 3 are referenced with the volume number followed by its page number (e.g., ***Abies*** 1–5; buttercup 3–24). Entries from this volume (Part 4) are followed by a page number only.

Note: This index is a compilation of the indices from Parts 1 through 4 of the Vascular Plants series. Taxa from Parts 1 through 3 are referenced with the volume number followed by its page number (e.g., ***Abies*** 1–5; buttercup 3–24). Entries from this volume (Part 4) are followed by a page number only.

Note: This index is a compilation of the indices from Parts 1 through 4 of the Vascular Plants series. Taxa from Parts 1 through 3 are referenced with the volume number followed by its page number (e.g., ***Abies*** 1–5; buttercup 3–24). Entries from this volume (Part 4) are followed by a page number only.

Note: This index is a compilation of the indices from Parts 1 through 4 of the Vascular Plants series. Taxa from Parts 1 through 3 are referenced with the volume number followed by its page number (e.g., ***Abies*** 1–5; buttercup 3–24). Entries from this volume (Part 4) are followed by a page number only.

Note: This index is a compilation of the indices from Parts 1 through 4 of the Vascular Plants series. Taxa from Parts 1 through 3 are referenced with the volume number followed by its page number (e.g., ***Abies*** 1–5; buttercup 3–24). Entries from this volume (Part 4) are followed by a page number only.

Note: This index is a compilation of the indices from Parts 1 through 4 of the Vascular Plants series. Taxa from Parts 1 through 3 are referenced with the volume number followed by its page number (e.g., ***Abies*** 1–5; buttercup 3–24). Entries from this volume (Part 4) are followed by a page number only.

Note: This index is a compilation of the indices from Parts 1 through 4 of the Vascular Plants series. Taxa from Parts 1 through 3 are referenced with the volume number followed by its page number (e.g., ***Ables*** 1–5; buttercup 3–24). Entries from this volume (Part 4) are followed by a page number only.

Note: This index is a compilation of the indices from Parts 1 through 4 of the Vascular Plants series. Taxa from Parts 1 through 3 are referenced with the volume number followed by its page number (e.g., ***Abies*** 1–5; buttercup 3–24). Entries from this volume (Part 4) are followed by a page number only.

Note: This index is a compilation of the indices from Parts 1 through 4 of the Vascular Plants series. Taxa from Parts 1 through 3 are referenced with the volume number followed by its page number (e.g., ***Abies*** 1–5; buttercup 3–24). Entries from this volume (Part 4) are followed by a page number only.

Note: This index is a compilation of the indices from Parts 1 through 4 of the Vascular Plants series. Taxa from Parts 1 through 3 are referenced with the volume number followed by its page number (e.g., ***Abies*** 1–5; buttercup 3–24). Entries from this volume (Part 4) are followed by a page number only.

Note: This index is a compilation of the indices from Parts 1 through 4 of the Vascular Plants series. Taxa from Parts 1 through 3 are referenced with the volume number followed by its page number (e.g., ***Abies*** 1–5; buttercup 3–24). Entries from this volume (Part 4) are followed by a page number only.

Note: This index is a compilation of the indices from Parts 1 through 4 of the Vascular Plants series. Taxa from Parts 1 through 3 are referenced with the volume number followed by its page number (e.g., ***Abies*** 1–5; buttercup 3–24). Entries from this volume (Part 4) are followed by a page number only.

Note: This index is a compilation of the indices from Parts 1 through 4 of the Vascular Plants series. Taxa from Parts 1 through 3 are referenced with the volume number followed by its page number (e.g., ***Abies*** 1–5; buttercup 3–24). Entries from this volume (Part 4) are followed by a page number only.

Note: This index is a compilation of the indices from Parts 1 through 4 of the Vascular Plants series. Taxa from Parts 1 through 3 are referenced with the volume number followed by its page number (e.g., ***Abies*** 1–5; buttercup 3–24). Entries from this volume (Part 4) are followed by a page number only.

Note: This index is a compilation of the indices from Parts 1 through 4 of the Vascular Plants series. Taxa from Parts 1 through 3 are referenced with the volume number followed by its page number (e.g., ***Abies*** 1–5; buttercup 3–24). Entries from this volume (Part 4) are followed by a page number only.

Note: This index is a compilation of the indices from Parts 1 through 4 of the Vascular Plants series. Taxa from Parts 1 through 3 are referenced with the volume number followed by its page number (e.g., ***Abies*** 1–5; buttercup 3–24). Entries from this volume (Part 4) are followed by a page number only.

Note: This index is a compilation of the indices from Parts 1 through 4 of the Vascular Plants series. Taxa from Parts 1 through 3 are referenced with the volume number followed by its page number (e.g., ***Abies*** 1–5; buttercup 3–24). Entries from this volume (Part 4) are followed by a page number only.

Note: This index is a compilation of the indices from Parts 1 through 4 of the Vascular Plants series. Taxa from Parts 1 through 3 are referenced with the volume number followed by its page number (e.g., ***Abies*** 1–5; buttercup 3–24). Entries from this volume (Part 4) are followed by a page number only.

Note: This index is a compilation of the indices from Parts 1 through 4 of the Vascular Plants series. Taxa from Parts 1 through 3 are referenced with the volume number followed by its page number (e.g., ***Abies*** 1–5; buttercup 3–24). Entries from this volume (Part 4) are followed by a page number only.

Note: This index is a compilation of the indices from Parts 1 through 4 of the Vascular Plants series. Taxa from Parts 1 through 3 are referenced with the volume number followed by its page number (e.g., ***Abies*** 1–5; buttercup 3–24). Entries from this volume (Part 4) are followed by a page number only.

Note: This index is a compilation of the indices from Parts 1 through 4 of the Vascular Plants series. Taxa from Parts 1 through 3 are referenced with the volume number followed by its page number (e.g., ***Abies*** 1–5; buttercup 3–24). Entries from this volume (Part 4) are followed by a page number only.

Note: This index is a compilation of the indices from Parts 1 through 4 of the Vascular Plants series. Taxa from Parts 1 through 3 are referenced with the volume number followed by its page number (e.g., ***Abies*** 1–5; buttercup 3–24). Entries from this volume (Part 4) are followed by a page number only.

Note: This index is a compilation of the indices from Parts 1 through 4 of the Vascular Plants series. Taxa from Parts 1 through 3 are referenced with the volume number followed by its page number (e.g., ***Abies*** 1–5; buttercup 3–24). Entries from this volume (Part 4) are followed by a page number only.

Note: This index is a compilation of the indices from Parts 1 through 4 of the Vascular Plants series. Taxa from Parts 1 through 3 are referenced with the volume number followed by its page number (e.g., ***Abies*** 1–5; buttercup 3–24). Entries from this volume (Part 4) are followed by a page number only.

Note: This index is a compilation of the indices from Parts 1 through 4 of the Vascular Plants series. Taxa from Parts 1 through 3 are referenced with the volume number followed by its page number (e.g., ***Abies*** 1–5; buttercup 3–24). Entries from this volume (Part 4) are followed by a page number only.

Note: This index is a compilation of the indices from Parts 1 through 4 of the Vascular Plants series. Taxa from Parts 1 through 3 are referenced with the volume number followed by its page number (e.g., ***Abies*** 1–5; buttercup 3–24). Entries from this volume (Part 4) are followed by a page number only.

Note: This index is a compilation of the indices from Parts 1 through 4 of the Vascular Plants series. Taxa from Parts 1 through 3 are referenced with the volume number followed by its page number (e.g., ***Abies*** 1–5; buttercup 3–24). Entries from this volume (Part 4) are followed by a page number only.

Note: This index is a compilation of the indices from Parts 1 through 4 of the Vascular Plants series. Taxa from Parts 1 through 3 are referenced with the volume number followed by its page number (e.g., ***Abies*** 1–5; buttercup 3–24). Entries from this volume (Part 4) are followed by a page number only.

Note: This index is a compilation of the indices from Parts 1 through 4 of the Vascular Plants series. Taxa from Parts 1 through 3 are referenced with the volume number followed by its page number (e.g., ***Abies*** 1–5; buttercup 3–24). Entries from this volume (Part 4) are followed by a page number only.

Note: This index is a compilation of the indices from Parts 1 through 4 of the Vascular Plants series. Taxa from Parts 1 through 3 are referenced with the volume number followed by its page number (e.g., ***Abies*** 1–5; buttercup 3–24). Entries from this volume (Part 4) are followed by a page number only.

Note: This index is a compilation of the indices from Parts 1 through 4 of the Vascular Plants series. Taxa from Parts 1 through 3 are referenced with the volume number followed by its page number (e.g., ***Abies*** 1–5; buttercup 3–24). Entries from this volume (Part 4) are followed by a page number only.

Note: This index is a compilation of the indices from Parts 1 through 4 of the Vascular Plants series. Taxa from Parts 1 through 3 are referenced with the volume number followed by its page number (e.g., ***Abies*** 1–5; buttercup 3–24). Entries from this volume (Part 4) are followed by a page number only.

Note: This index is a compilation of the indices from Parts 1 through 4 of the Vascular Plants series. Taxa from Parts 1 through 3 are referenced with the volume number followed by its page number (e.g., ***Abies*** 1–5; buttercup 3–24). Entries from this volume (Part 4) are followed by a page number only.

Note: This index is a compilation of the indices from Parts 1 through 4 of the Vascular Plants series. Taxa from Parts 1 through 3 are referenced with the volume number followed by its page number (e.g., ***Abies*** 1–5; buttercup 3–24). Entries from this volume (Part 4) are followed by a page number only.

Note: This index is a compilation of the indices from Parts 1 through 4 of the Vascular Plants series. Taxa from Parts 1 through 3 are referenced with the volume number followed by its page number (e.g., ***Abies*** 1–5; buttercup 3–24). Entries from this volume (Part 4) are followed by a page number only.

Note: This index is a compilation of the indices from Parts 1 through 4 of the Vascular Plants series. Taxa from Parts 1 through 3 are referenced with the volume number followed by its page number (e.g., ***Abies*** 1–5; buttercup 3–24). Entries from this volume (Part 4) are followed by a page number only.

Note: This index is a compilation of the indices from Parts 1 through 4 of the Vascular Plants series. Taxa from Parts 1 through 3 are referenced with the volume number followed by its page number (e.g., ***Abies*** 1–5; buttercup 3–24). Entries from this volume (Part 4) are followed by a page number only.

Note: This index is a compilation of the indices from Parts 1 through 4 of the Vascular Plants series. Taxa from Parts 1 through 3 are referenced with the volume number followed by its page number (e.g., ***Abies*** 1–5; buttercup 3–24). Entries from this volume (Part 4) are followed by a page number only.

Note: This index is a compilation of the indices from Parts 1 through 4 of the Vascular Plants series. Taxa from Parts 1 through 3 are referenced with the volume number followed by its page number (e.g., ***Abies*** 1–5; buttercup 3–24). Entries from this volume (Part 4) are followed by a page number only.

Note: This index is a compilation of the indices from Parts 1 through 4 of the Vascular Plants series. Taxa from Parts 1 through 3 are referenced with the volume number followed by its page number (e.g., ***Abies*** 1–5; buttercup 3–24). Entries from this volume (Part 4) are followed by a page number only.

Note: This index is a compilation of the indices from Parts 1 through 4 of the Vascular Plants series. Taxa from Parts 1 through 3 are referenced with the volume number followed by its page number (e.g., ***Abies*** 1–5; buttercup 3–24). Entries from this volume (Part 4) are followed by a page number only.

Note: This index is a compilation of the indices from Parts 1 through 4 of the Vascular Plants series. Taxa from Parts 1 through 3 are referenced with the volume number followed by its page number (e.g., ***Abies*** 1–5; buttercup 3–24). Entries from this volume (Part 4) are followed by a page number only.

Note: This index is a compilation of the indices from Parts 1 through 4 of the Vascular Plants series. Taxa from Parts 1 through 3 are referenced with the volume number followed by its page number (e.g., ***Abies*** 1–5; buttercup 3–24). Entries from this volume (Part 4) are followed by a page number only.

Note: This index is a compilation of the indices from Parts 1 through 4 of the Vascular Plants series. Taxa from Parts 1 through 3 are referenced with the volume number followed by its page number (e.g., ***Abies*** 1–5; buttercup 3–24). Entries from this volume (Part 4) are followed by a page number only.

ERRATA AND ADDENDA
THE VASCULAR PLANTS OF BRITISH COLUMBIA
PARTS 1, 2, 3 AND 4

PART 1 – GYMNOSPERMS AND DICOTYLEDONS (ACERACEAE THROUGH CUCURBITACEAE)*

Page 16 – Line 10 down – add:

***Cnidium cnidiifolium* (Turcz.) Schischk.** (*Conioselinum cnidiifolium* [Turcz.] Porsild)

Hemlock-parsley

Habitat/Range: Mesic to dry forest openings, meadows and roadsides in the montane zone; rare in NW BC, known only from the Atlin Road; amphiberingian, N to AK, YT and NT; E Asia.

Notes: The first BC collection of this species was made by Frank Lomer in 1992. It will key out to *Lomatium* (lead 31, page 14) but differs in being distinctly caulescent and having a more northern range than any of our *Lomatium* species.

Page 34 – line 12 from the bottom – add:

***Agoseris lackschewitzii* Henderson & Mosely**

Pink agoseris

Habitat/Range: Moist to mesic meadows in the montane to subalpine zone; rare in SE BC, known only from Yoho National Park; E to SW AB and S to NW MT and N ID.

Notes: This distinctive species, recently described by Henderson and Mosely (1990), Yoho National Park by Peter Achuff in 1991. It may be distinguished from our other *Agoseris* species by its pink rays and achenes that gradually taper into beaks 4.2–6.6 mm long.

Page 50 – last line – add:

***Brickellia grandiflora* (Hook.) Nutt.**

Large-flowered brickellia

Habitat/Range: Mesic to dry gravelly slopes and rock outcrops in the montane zone; rare in SE BC, known only from the Flathead Valley; S to NB, MO, AR and NM.

Notes: The first BC record for this species was collected in lower Flathead Valley in 1993 by Frank Lomer. It is easily distinguished from our other entire, linear to oblong-leaved *Brickellia* species (*B. oblongifolia*) by its toothed, deltoid-ovate to subcordate leaves.

Page 53 – line 16 from the bottom – add:

***Chondrilla juncea* L.**

Gum-succory

Habitat/Range: Dry roadsides and waste areas; rare in SC BC; introduced from Eurasia.

Notes: First reported for BC by Ernie McNaughton in 1990, this species with its milky stem juices, will key out to *Taraxacum* (lead 6, page 26). It differs from the latter in having a branching, rushlike habit, several to numerous small heads and only a few (usually about 11) ray flowers.

Page 63 – line 9 down – *Erigeron polispermus* var. *polispermus* should read *Erigeron poliospermus* var. *poliospermus*

---

* Some of these species were recently collected or discovered in herbaria during 1993.

Page 65 – line 5 down – add:

Galinsoga parviflora **Cav.**

Small-flowered galinsoga

Habitat/Range: Disturbed sites in the steppe zone; rare in SC BC, known recently only from Vaseux Lake; introducted from Mexico and South America.

Notes: Collected in 1993 by Malcolm Martin and Ernie McNaughton. It may be distinguished from *Galinsoga ciliata* by its less toothed leaves, a reduced or absent ray pappus and disk pappus with fimbriate and more or less blunt scales.

Page 69 – line 16 down – add:

Aphragmus eschscholtzianus **Andrz. *ex* DC.**

Eschscholtz's little nightmare

Habitat/Range: Moist to wet solifluction slopes and rocky ridges in the alpine zone; rare in NW BC, amphiberingian, N to AK and YT; NE Asia.

Notes: Collected in NW BC in 1992 by D. Meidinger. It represents a southern extension of range in Canada from just north of the BC border in the Kluane Ranges, YT (Douglas and Ruyle-Douglas 1978). It will key out to *Arabis* (lead 68) in the Brassicaceae key but differs in having entire leaves.

Page 94 – line 15 down – add:

Amsinckia retrorsa **Suksd.**

Rigid fiddleneck

Habitat/Range: Dry roadsides, fields and disturbed sites in the steppe zone; rare in SC BC, known only from the Kelowna area; S to UT and CA.

Notes: The first BC record for this species was recently taken by Malcolm Martin and Ernie McNaughton. It is distinguished from *Amsinckia intermedia* and *A. menziesii* by having spreading-hispid hairs with shorter, softer, more or less retrorse hairs throughout the stems.

Page 95 – line 8 from bottom – add:

Cryptantha watsonii **(A. Gray) Greene**

Watson's cryptantha

Habitat/Range: Dry sites in the montane zone; rare in SC BC, known only from Osoyoos and the Ashnola River; S to MT, CO and NV.

Notes: First collected at Osoyoos in 1940 by J. Eastam, but misidentified. The latter specimen was reidentified by Frank Lomer, who also collected a second record from along the Ashnola River in 1993. This species will key out to *C. fendleri* (lead 5, page 95) but differs from the latter in having sharply edged nutlets.

Page 128 – *Thelypodium* – After examining numerous sheets of *Thelypodium laciniatum* and *T. milleflorum* we find that, at least in our range, the two are often scarcely distinguishable. We therefore recommend that the treatment by Hitchcock and Cronquist (1973), who recognized these two taxa as varieties of *T. laciniatum*, be followed.

Page 144 – line 2 down, after ***PETRORHAGIA*** – add:

1. Flowers solitary or sometimes fasciculate .... *P. prolifera*
1. Flowers capitate .... *P. saxifraga*

Petrorhagia prolifera **(L.) P.W. Ball & Heywood** (*Tunica prolifera* [L.] Scop.)
Petrorhagia
Habitat/Range: Sandy, open sites in the lowland zone; rare in SW BC, known only from Comox; introduced from Europe.
Notes: This species was first collected by B. Bennett in 1992.

Page 155 – line 13 from bottom – add:

Atriplex longipes **Dreje**
Long-stalked orache
Habitat/Range: Waste places in the lowland zone; known only from Surrey and White Rock; introduced from Eurasia.
Notes: This species was collected recently by Frank Lomer. It will key out to *A. patula*, but the long-stalked, recurved fruits are distinctive.

Atriplex nuttallii **S. Wats.**
Nuttall's orache
Habitat/Range: Dry slopes in the steppe zone; rare in NE BC, known only from Clayhurst Crossing; E to SK and S to NM, AZ and NW CA.
Notes: This species is a dioecious sub-shrub (woody at least at the base). None of our other species of *Atriplex* has these two characters.

Page 158 – line 12 from the bottom – add:

POLYCNEMUM

Polycnemum arvense **L.**
Crunch weed
Habitat/Range: Dry roadsides and forest edges; rare in SC BC; introduced from Eurasia.
Notes: This species was first collected in BC by Malcolm Martin above Okanagan Lake. It will key out to *Salsola* (lead 2, page 153) but has perianth segments without appendages and elongate fruits while *Salsola* has winged perianth segments and cup-shaped fruits.

PART 2 – DICOTYLEDONS (DIAPENSIACEAE THROUGH PORTULACACEAE)

Page 15 – line 13 – add:

PLEURICOSPORA

Pleuricospora fimbriolata **A. Gray**
This species is known from a single collection taken in 1916 at Horne Lake, Vancouver Island (Wallace 1975). It is easily mistaken for *Hemitomes congestum* at first glance but the fimbriate perianths and the fimbriate and glabrous bracts readily distinguish it. Szczawinski (1962) identified this collection as the latter, thus *Pleuricospora fimbriolata* did not appear in his Ericaceae treatment for BC.

Page 26 – line 16 down – add:

***Astragalus drummondii* Dougl. ex Hook.**

Drummond's milk-vetch

Habitat/Range: Dry, open slopes in the montane zone; rare in SW BC, known only from the Sparwood area; E to SK and S to KS, UT and NM.

Notes: Although listed by Taylor and MacBryde (1977), no material was discovered in herberia until recently. This species will not key out easily in the present key. It is characterized by tall (4–8 dm), erect grayish-villous-hirsute stems and stipitate, glabrous, 2-locular pods. None of our other *Astragalus* species have this combination of characters.

Page 27 – line 23 down – add:

Astragalus spaldingii* A. Gray var. *spaldingii

Spalding's milk-vetch

Habitat/Range: Dry sites in the steppe vegetation zone; rare in SC BC, known only from Osoyoos; S to ID and NE OR.

Notes: This taxon was discovered in 1990 by Trevor Goward. The plants have white flowers (the banners bluish-lined) and pods that are small (4–7 mm long), sessile, ovoid and densely strigose-villous. These plants will key out to lead 22 but will go no farther since the pods are shorter than those given in either option.

Page 76 – line 15 down – add:

ABUTILON

***Abutilon theophrasti* Medic.**

Velvet-leaf

Habitat/Range: Cultivated fields and waste places in the lowland zone; rare in SW BC, known recently from Chilliwack and Abbotsford; introduced from Asia.

Notes: Until recently (1991), this species was known only from a 1944 report from Lumby (Scoggan 1978). This species will key out to lead 2 in the generic key but differs from *Iliamna* and *Sphaeralcea* in having entire or remotely crenate leaves rather than lobed ones.

Page 94 – line 17 down – add:

***Papaver rhoeas* L.**

Corn poppy

Habitat/Range: Waste places in the lowland and montane zone; rare garden escape in SW, SC and SE BC; introduced from Eurasia.

Notes: Collections of this species are known from Cadboro Bay and Fort Steele. This species will key out to *P. somniferum* but the brick-red to scarlet or deep purple flowers and deeply pinnatifid leaves readily distinguish it from the latter.

Page 109 – line 12 from the bottom – add:

***Polygonum minus* Huds.**

Small knotweed

Habitat/Range: Moist to mesic waste places in the lowland and steppe zones; infrequent in SW and SC BC; introduced from Europe.

Notes: This species has been collected at several localities in the lower Fraser Valley and at Tranquille. It will key out to *P. hydropiperoides*, but this species is a tap-rooted annual.

Page 111 – *Rumex* key – the 10 leads should read:

10a. Leaves more or less truncate or cordate at the base; valves usually cordate .... *R. patientia*

10a. Leaves rounded to acute at the base; valves usually triangular-ovate or oblong.

10b. Panicles dense, the verticils of the flowers often contiguous, at least towards the branch ends; leaves strongly crisp-margined .... *R. crispus*

10b. Panicles open, the verticils of the flowers not contiguous; leaves not strongly crisp-margined .... *R. conglomeratus*

Page 112 – line 10 from bottom – add:

***Rumex patientia* L.**

Patience dock

Habitat/Range: Roadsides and waste places in the lowland and montane zones; rare in SW and SC BC; introduced from Europe.

Notes: This species has been collected, independently, several times in recent years by both F. Lomer and M. Martin. It was first collected almost 50 years ago by J.W. Eastham but the collection remained misidentified until recently.

PART 3 – DICOTYLEDONS (PRIMULACEAE THROUGH ZYGOPHYLLACEAE AND PTERIDOPHYTES)

Page 70 – line 7 – add:

***Chrysosplenium wrightii* Franc. & Savat**

Wright's golden-saxifrage or golden carpet

Habitat/Range: Wet seepage sites, solifluction slopes and scree in the alpine zone; rare in NW BC; amphiberingian, N to AK, YT and NT; NE Asia.

Notes: This species was collected in NW BC by Jim Pojar and Del Meidinger in 1992. It represents a southern extension of range in Canada from about 62°N in the YT. It is easily distinguished from *C. tetrandrum*, which can also have 8 stamens, by its purple-mottled sepals and reddish-pubescent petioles.

Page 75 – leads 24 – species are reversed.

Page 76 – line 20 down – add:

***Saxifraga davurica* Willd. ssp. *grandipetala* (Engl. & Irmsch.) Hulten**

Large-petalled saxifrage

Habitat/Range: Moist to wet solifluction slopes, scree and seepage sites in the alpine zone; rare in NW BC; amphiberingian, N to AK, YT and NT; NE Asia.

Notes: Collected in NW BC in 1992 by Jim Pojar and Del Meidinger. This taxon was originally mapped for NW BC by Hulten (1968) but no material was found in Canadian herbaria, thus we treated it as an excluded species. It will key out to *Saxifraga lyalli* but differs in having subulate to linear staminal filaments and more narrow leaves with longer petioles.

Page 177 – line 4 down – *Floerka proserpinacoides* should read *Floerkea proserpinacoides.* This species was recently collected for the second time in BC by F. Lomer at Mt. Kobau in SC BC.

PART 4 – MONOCOTYLEDONS

Page 100 – line 21 down – add:

***Briza minor* L.**

Little quaking grass

Habitat/Range: Dry disturbed sites in the lowland zone; rare in SW BC, known recently only from Rocky Point on S Vancouver Island; horticultural species introduced from Europe.

Notes: This species was collected in 1993 by Mike Ryan at Rocky Point. Previously, it was known only from a few historic collections and not thought to persist.

Page 108 – line 11 down – add:

SIEGLINGIA

***Danthonia decumbens* (L.) DC.** (*Sieglingia decumbens* (L.) Bernh.)

Heather-grass

Habitat/Range: Moist to wet sites in the lowland zone; rare in SW BC, known only from Prevost Island; introduced from Europe.

Notes: This species was first collected in BC in 1972 by Hans Roemer. In the generic key it will key out to Group IV, lead 6 (page 90). It differs from *Koeleria* and *Sphenopholus* in having glumes and lemmas which are rounded and glabrous on the backs and from other *Danthonia* species by the absence of awns on the lemmas.

Page 115 – line 5 down – add:

EREMOPYRUM

***Eremopyrum triticeum* (Gaertn.) Nevski** (*Agropyron triticeum* Gaertn.)

Annual wheatgrass

Habitat/Range: Alkali meadows in the steppe zone; rare in SC BC, known only from Osoyoos; introduced from Eurasia.

Notes: Recently collected by Frank Lomer, this species resembles *A. cristatum*, but is annual and has shorter spikes (less than 1.5 cm long). In the generic key it will key out to *Agropyron.*

Page 123 – line 19 down – add:

***Leymus triticoides* (Buckl.) Pilger** (*Elymus triticoides* Buckl.)

Creeping wildrye

Habitat/Range: Sand dunes in the lowland zone; rare in SW BC, known only from Island View Beach, S Vancouver Island; S to CO, NM and MX.

Notes: The first BC record for this species was made in 1993 by Hans Roemer. It is surprising that it was not found previously since it is presumed to be one of the parents (with *L. mollis*) of the hybrid, *L.* x *vancouveriensis*, which occurs infrequently in SW BC. *Leymus triticoides* will key out to *L. cinereus* (lead 2, page 122) but differs from the latter in having shorter ligules (usually 0.5 mm long versus [2] 3–7 mm long) and narrower leaves (6 mm wide, or less versus [6] 10–20 mm wide).

Page 138 – top of page – add:

Poa pseudoabbreviata **Roshev.**

Polar bluegrass

Habitat/Range: Dry, open ridges from the montane to alpine zones; rare in NW BC, known only from the Horseranch Range; amphiberingian, N to AK, YT and NT, NE Asia.

Notes: This species was first collected in 1980 by C.J. Clements but was identified only recently by R.J. Soreng.